Pre-GED Connection™

WITHDRAWN

Mathematics

Instruction by Cathy Fillmore Hoyt

Reviews and Skill Practice
by Karen Lassiter

LiteracyLink® is a joint project of PBS,
Kentucky Educational Television,
the National Center on Adult Literacy,
and the Kentucky Department of Education.

This project is funded in whole,
or in part, by the Star Schools Program
of the USDE under contract #R203D60001.

 **PBS** LiteracyLink®   **KET**  **NCAL**

Acknowledgments

LiteracyLink® Advisory Board
Lynn Allen, Idaho Public Television
Anthony Buttino, WNED-TV
Anthony Carnevale, Educational
    Testing Service
Andy Chaves, Marriott International, Inc.
Patricia Edwards, Michigan State University
Phyllis Eisen, Center for Workforce Success National
    Association of Manufacturers
Maggi Gaines, Baltimore Reads, Inc.
Marshall Goldberg, Association of Joint Labor
    Management Educational Programs
Milton Goldberg, National Alliance
    for Business
Neal Johnson, Association of Governing Boards of
    Universities and Colleges
Cynthia Johnston, Central Piedmont Community
    College
Sandra Kestner, Kentucky Department for Adult
    Education and Literacy
Thomas Kinney, American Association of Adult and
    Continuing Education
Dale Lipschultz, American Library Association
Lennox McLendon, National Adult Education
    Professional Development Consortium
Cam Messina, KLRN
Patricia Miller, KNPB
Cathy Powers, WLRN
Ray Ramirez, U.S. Department of Education
Emma Rhodes, (retired) Arkansas Department of
    Education
Cynthia Ruiz, KCET
Tony Sarmiento, Worker Centered Learning,
    Working for America Institute
Steve Steurer, Correctional
    Education Association
LaShell Stevens-Staley, Iowa PTV
Fran Tracy-Mumford, Delaware Department of
    Adult/Community Education
Terilyn Turner, Community Education,
    St. Paul Public Schools

**LiteracyLink®
Ex Officio Advisory Board**
Joan Auchter, GED Testing Service
Barbara Derwart, U.S. Department of Labor
Cheryl Garnette, OERI, U.S.
    Department of Education
Andrew Hartman, National Institute
    for Literacy
Mary Lovell, OVAE, U.S. Department
    of Education
Ronald Pugsley, OVAE, U.S. Department
    of Education
Linda Roberts, U.S. Department of Education
Joe Wilkes, OERI, U.S. Department of Education

LiteracyLink® Partners
LiteracyLink® is a joint project of:
    Public Broadcasting Service,
    Kentucky Educational Television,
    National Center on Adult Literacy, and the
    Kentucky Department of Education.

**Content Design and Workbook
Editorial Development**
    Learning Unlimited, Oak Park, Illinois
Design and Layout
    By Design, Lexington, Kentucky
Project Coordinators
    Milli Fazey, KET, Lexington, Kentucky
    Margaret Norman, KET, Lexington, Kentucky

This project is funded in whole, or in part, by the
Star Schools Program of the USDE under contract
#R203D60001.

PBS LiteracyLink® is a registered mark of the
Public Broadcasting Service.

Contents

Introduction

Welcome to *Pre-GED Mathematics*. This workbook is part of the *LiteracyLink®* multimedia educational system for adult learners and educators. The system includes *Pre-GED Connection*, which builds a foundation for GED-level study and *GED Connection*, which learners use to study for the GED Tests. *LiteracyLink* also includes *Workplace Essential Skills*, which targets upgrading the knowledge and skills needed to succeed in the world of work.

Instructional Programs

Pre-GED Connection consists of 26 instructional video programs and five companion workbooks. Each *Pre-GED Connection* workbook lesson accompanies a video program. For example, the first lesson in this book is *Program 20— Number Sense*. This workbook lesson should be used with *Pre-GED Connection Video Program 20—Number Sense*. In addition, you can go online to www.pbs.org/literacy and click the *Pre-GED Math* link.

Who's Responsible for LiteracyLink®?

LiteracyLink was developed through a five-year grant by the U.S. Department of Education. The following partners have contributed to the development of the *LiteracyLink* system:

| PBS Adult Learning Service | Kentucky Educational Television (KET) | The National Center on Adult Literacy (NCAL) of the University of Pennsylvania | The Kentucky Department of Education |

All of the *LiteracyLink* partners wish you the very best in meeting all of your educational goals.

Pre-GED CONNECTION consists of these educational tools:

26 VIDEO PROGRAMS shown on public television and in adult learning centers

ONLINE MATERIALS available on the Internet at http://www.pbs.org/literacy

FIVE Pre-GED COMPANION WORKBOOKS
Language Arts, Writing
Language Arts, Reading
Social Studies
Science
Mathematics

Getting Started with *Pre-GED Connection Mathematics*

Before you start using the workbook, take some time to preview its features.

1. Take the **Pretest** on page 6. This will help you decide which areas you need to focus on. You should use the evaluation chart on page 17 to develop your study plan.

2. Work through the **workbook lessons**—each one corresponds to a video program.

   The *Before You Watch* feature sets up the video program:
   - **Think About the Topic** gives a brief overview of the video
   - **Prepare to Watch the Video** is a short activity with instant feedback that shows how everyday knowledge can help you better understand the topic
   - **Lesson Goals** highlight the main ideas of each video and workbook lesson
   - **Terms** introduces key math vocabulary

   The *After You Watch* feature helps you evaluate what you saw:
   - **Think About the Program** presents questions that focus on key points from the video
   - **Make the Connection** applies what you have learned to real-life situations

   Three *Math Skills* sections correspond to key concepts in the video program.

   The *GED Problem Solving* feature introduces the types of problem-solving skills that you will see on the GED.

   The *GED Math Connection* introduces you to the calculator, grids, and formulas page that you will see on the GED.

   GED Math Practice allows you to practice with the types of problems that you will see on the actual test.

3. Take the **Posttest** on page 188 to determine your progress and whether you are ready for GED-level work.

4. Use the **Answer Key** to check your answers.

5. Refer to the **Math Handbook** at the back of the book as needed.

6. Use the **Extra Practice** at the back of the book for additional practice.

For Teachers

Portions of *LiteracyLink* have been developed for adult educators and service providers. Teachers can use Pre-GED lesson plans in the *LiteracyLink Teacher's Guide* binder. This binder also contains lesson plans for *GED Connection* and *Workplace Essential Skills*.

Math Pretest

Directions

The Math Pretest has 25 questions to measure your math and problem-solving skills. After you complete the test, check your answers on pages 17–18. Then use the evaluation chart on page 19 to identify the math skills that you need to work on.

The Pretest consists of Part I and Part II. You will be allowed to use a calculator on Part I, but you are not required to use it. You <u>may not</u> use a calculator on Part II of the Pretest.

Most questions on the Math Pretest are multiple choice with five answer choices. A few questions do not have choices. For these questions, you should work the problem and then fill in your answer on the special answer grid on your answer sheet. Study the sample questions below to see how to use the answer sheet.

EXAMPLE: Sandra has $50 in the bank. If she writes a check for $15.84, how much will she have left in her account?

(1) $35.26

(2) $35.16

(3) $34.26

(4) $34.16

(5) Not enough information is given.

① ② ③ ④ ⑤

Answer: The correct answer is **(4) $34.16.** Fill in answer space 4 on the answer sheet.

EXAMPLE: Leo has two bottles of floor wax. One bottle holds $\frac{3}{4}$ gallon, and the other holds $\frac{1}{2}$ gallon. How many gallons of floor wax does he have?

Answer: Leo has **$1\frac{1}{4}$ gallons** of floor wax. Examples of how the answer could be recorded correctly are shown below.

Remember these rules:

- Fill in only one circle in each column.
- You can start in any column as long as the answer fits within the grid.
- Mixed numbers such as $1\frac{1}{4}$ must be recorded as a decimal (1.25) or an improper fraction $\frac{5}{4}$.

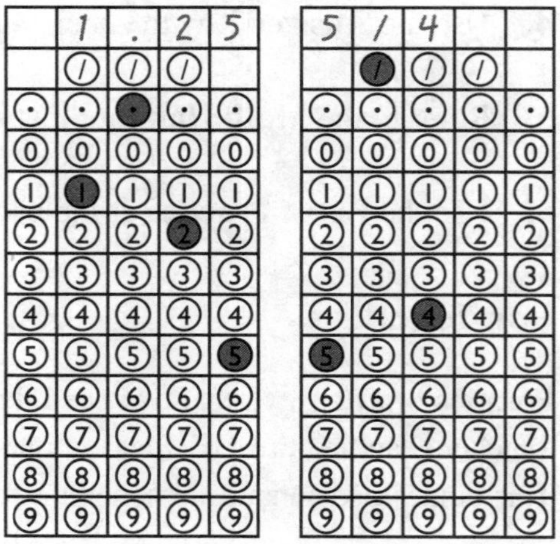

Pretest Answer Sheet

Part I

1. ① ② ③ ④ ⑤
2. ① ② ③ ④ ⑤
3. ① ② ③ ④ ⑤
4. ① ② ③ ④ ⑤
5. ① ② ③ ④ ⑤

6.
	/	/	/	
·	·	·	·	·
0	0	0	0	0
1	1	1	1	1
2	2	2	2	2
3	3	3	3	3
4	4	4	4	4
5	5	5	5	5
6	6	6	6	6
7	7	7	7	7
8	8	8	8	8
9	9	9	9	9

7. ① ② ③ ④ ⑤
8. ① ② ③ ④ ⑤
9. ① ② ③ ④ ⑤

10.
	/	/	/	
·	·	·	·	·
0	0	0	0	0
1	1	1	1	1
2	2	2	2	2
3	3	3	3	3
4	4	4	4	4
5	5	5	5	5
6	6	6	6	6
7	7	7	7	7
8	8	8	8	8
9	9	9	9	9

11. ① ② ③ ④ ⑤
12. ① ② ③ ④ ⑤
13. ① ② ③ ④ ⑤

Part II

14. ① ② ③ ④ ⑤
15. ① ② ③ ④ ⑤

16.
	/	/	/	
·	·	·	·	·
0	0	0	0	0
1	1	1	1	1
2	2	2	2	2
3	3	3	3	3
4	4	4	4	4
5	5	5	5	5
6	6	6	6	6
7	7	7	7	7
8	8	8	8	8
9	9	9	9	9

17. ① ② ③ ④ ⑤
18. ① ② ③ ④ ⑤
19. ① ② ③ ④ ⑤

20.
	/	/	/	
·	·	·	·	·
0	0	0	0	0
1	1	1	1	1
2	2	2	2	2
3	3	3	3	3
4	4	4	4	4
5	5	5	5	5
6	6	6	6	6
7	7	7	7	7
8	8	8	8	8
9	9	9	9	9

21. ① ② ③ ④ ⑤
22. ① ② ③ ④ ⑤
23. ① ② ③ ④ ⑤
24. ① ② ③ ④ ⑤
25. ① ② ③ ④ ⑤

FORMULAS

AREA of a:

square	Area $=$ side^2
rectangle	Area $=$ length $\times$ width
parallelogram	Area $=$ base $\times$ height
triangle	Area $= \frac{1}{2} \times$ base $\times$ height
trapezoid	Area $= \frac{1}{2} \times$ (base$_1$ + base$_2$) $\times$ height
circle	Area $= \pi \times$ radius^2; π is approximately equal to 3.14.

PERIMETER of a:

square	Perimeter $= 4 \times$ side
rectangle	Perimeter $= 2 \times$ length $+ 2 \times$ width
triangle	Perimeter $=$ side$_1$ + side$_2$ + side$_3$

CIRCUMFERENCE of a:

circle	Circumference $= \pi \times$ diameter;
	π is approximately equal to 3.14.

VOLUME of a:

cube	Volume $=$ edge^3
rectangular solid	Volume $=$ length $\times$ width $\times$ height
square pyramid	Volume $= \frac{1}{3} \times$ (base edge)^2 $\times$ height
cylinder	Volume $= \pi \times$ radius^2 $\times$ height;
	π is approximately equal to 3.14.
cone	Volume $= \frac{1}{3} \times \pi \times$ radius^2 $\times$ height;
	π is approximately equal to 3.14.

COORDINATE GEOMETRY

distance between points $= \sqrt{(x_2 - x_1)^2 + (y_2 - y_1)^2}$; (x_1, y_1) and (x_2, y_2) are two points in a plane.

slope of a line $= \frac{y_2 - y_1}{x_2 - x_1}$; (x_1, y_1) and (x_2, y_2) are two points on the line.

PYTHAGOREAN RELATIONSHIP

$a^2 + b^2 = c^2$; a and b are legs and c the hypotenuse of a right triangle.

MEASURES OF CENTRAL TENDENCY

mean $= \frac{x_1 + x_2 + ... + x_n}{n}$, where the x's are the values for which a mean is desired, and n is the total number of values for x.

median $=$ the middle value of an odd number of <u>ordered</u> scores, and halfway between the two middle values of an even number of <u>ordered</u> scores.

SIMPLE INTEREST　　　interest $=$ principal $\times$ rate $\times$ time

DISTANCE　　　distance $=$ rate $\times$ time

TOTAL COST　　　total cost $=$ (number of units) $\times$ (price per unit)

Math Pretest

Part I

Write your answers on the answer sheet provided on page 7. You may use a calculator for the items in Part I, but some of the items may be solved more quickly without a calculator.

Some of the questions will require you to use a formula. The formulas you need are given on page 8. Not all of the formulas on that page will be needed.

Choose the <u>one best answer</u> to each question.

<u>Question 1</u> refers to the following information.

$50 OFF
C O U P O N
19-inch
flat screen monitor
Regular Price $469
Shop & Save at
PC Central

1. Frank saw the same monitor at Big Buys for $480. If Frank uses the coupon, how much less will he pay for the monitor at PC Central than at Big Buys?
   - **(1)** $11
   - **(2)** $61
   - **(3)** $70
   - **(4)** $120
   - **(5)** $419

2. Tom works at Silk Screen T's. He can print 60 T-shirts per hour. If he starts at 8:30 A.M., when will he finish an order of 240 T-shirts?
   - **(1)** 11:30 A.M.
   - **(2)** 12 noon
   - **(3)** 12:30 P.M.
   - **(4)** 1 P.M.
   - **(5)** 1:30 P.M.

3. Judy is planning a luncheon for 196 workers and 28 members of the management team. The caterer for the event charges $9 per person. How much will Judy pay the caterer?
   - **(1)** $224
   - **(2)** $233
   - **(3)** $1,512
   - **(4)** $1,764
   - **(5)** $2,016

Questions 4 and 5 refer to the following information.

Easy-to-Sew Patterns Fabric List	
SHIRT	
If the fabric width is...	You need...
45 inches	$1\frac{7}{8}$ yards
60 inches	$1\frac{1}{4}$ yards
PANTS	
If the fabric width is...	You need...
45 inches	$2\frac{3}{8}$ yards
60 inches	$1\frac{1}{2}$ yards

4. Jenna is using the pattern above to make her daughter three pairs of pants. If she uses a fabric with a 45-inch width, how many yards of fabric will she need?

(1) $7\frac{1}{8}$

(2) $6\frac{3}{4}$

(3) $6\frac{3}{8}$

(4) $4\frac{1}{2}$

(5) $3\frac{7}{8}$

5. A bolt of fabric has 40 yards. Exactly how many shirts can be made from the fabric on the bolt?

(1) 21

(2) 27

(3) 32

(4) 66

(5) Not enough information is given.

6. A small plane uses 6.8 gallons of fuel per hour. How many gallons of fuel will the plane use in 2.5 hours?

Mark your answer in the circles in the grid on your answer sheet.

Question 7 refers to the following information.

Kimball's
COPY CENTER
Small Business Special
15% Off
ANY Purchase or Service

7. Lansing had 3,000 copies made at Kimball's Copy Center. The regular rate for the copies is $0.04 per copy. If she used the coupon above, how much did she pay for her copies?
   (1) $18
   (2) $102
   (3) $105
   (4) $118
   (5) $120

8. A community center offers 16 classes for seniors and 24 classes for youth. What is the ratio of classes for seniors to the total number of classes?
   (1) 1:2
   (2) 1:3
   (3) 2:3
   (4) 2:5
   (5) 3:5

9. The regular price of a suit is $250. The suit was on sale for 20% off the regular price. When it did not sell, the manager took 10% off the sale price. What is the current price of the suit?
   (1) $175
   (2) $180
   (3) $200
   (4) $220
   (5) $225

10. Martin played a computer game four times and scored 10,000, 8,500, 12,500, and 9,500. What was Martin's mean score?

Mark your answer in the circles in the grid on your answer sheet.

<u>Questions 11 and 12</u> refer to the following information and graph.

Linda sells sandwiches in an office building. She made the following graph to show her sales for one week.

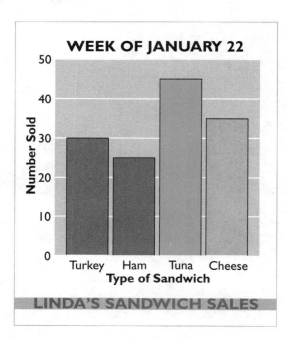

WEEK OF JANUARY 22

LINDA'S SANDWICH SALES

11. How many tuna sandwiches and cheese sandwiches were sold in all?
- **(1)** 10
- **(2)** 35
- **(3)** 45
- **(4)** 80
- **(5)** 135

12. If the information in the graph represents a typical week, how many ham sandwiches could Linda expect to sell over a 12-week period?
- **(1)** 300
- **(2)** 360
- **(3)** 420
- **(4)** 540
- **(5)** Not enough information is given.

13. A painting measures 22 inches by 26 inches. How many FEET of framing will be needed to build a frame?
- **(1)** 4.0
- **(2)** 4.8
- **(3)** 5.8
- **(4)** 8.0
- **(5)** 9.6

Part II

You may <u>not</u> use a calculator for the questions in Part II. Some of the questions will require you to use a formula. The formulas you need are given on page 8. Not all of the formulas on that page will be needed.

Choose the <u>one best answer</u> to each question.

14. The population of Fillmore is 48,164. Which of the following represents the population of Fillmore rounded to the NEAREST TEN THOUSAND?
- **(1)** 48,200
- **(2)** 47,000
- **(3)** 48,000
- **(4)** 49,000
- **(5)** 50,000

15. Paul and Sandy recently bought a new refrigerator. They agreed to pay $320 as a down payment and $62 per month for 18 months. Approximately how much will they pay in all for the refrigerator?
- **(1)** between $800 and $1,000
- **(2)** between $1,000 and $1,200
- **(3)** between $1,200 and $1,400
- **(4)** between $1,400 and $1,600
- **(5)** between $1,600 and $1,800

<u>Question 16</u> refers to the following map.

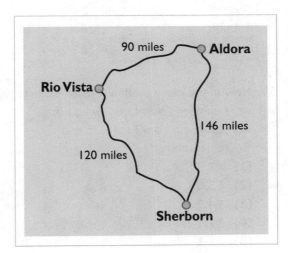

16. Carlos drives a delivery truck for a store in Rio Vista. Two days a week, he drives to Aldora and back. Three days a week he drives to Sherborn and back. How many miles does he drive per week?

Mark your answer in the circles in the grid on your answer sheet.

Questions 17 and 18 refer to the following table.

Question 19 refers to the following information.

ANNUAL GOODS SHIPPED (in millions of tons)	
Port	**Exports**
Cleveland	12.7
Boston	9.4
Freeport	5.3
St. Louis	28.2
Long Beach	22.4

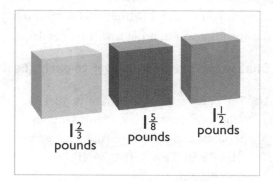

17. How many more million tons of exports were shipped from Long Beach than were shipped from Freeport and Boston combined?

(1) 7.7
(2) 13.0
(3) 14.7
(4) 17.1
(5) 37.1

18. The exports shipped from St. Louis are how many times the exports shipped from Boston?

(1) 2
(2) 3
(3) 4
(4) 5
(5) Not enough information is given.

19. Which of the following represents the weights of the packages arranged from least to greatest?

(1) $1\frac{1}{2}, 1\frac{2}{3}, 1\frac{5}{8}$
(2) $1\frac{5}{8}, 1\frac{1}{2}, 1\frac{2}{3}$
(3) $1\frac{1}{2}, 1\frac{5}{8}, 1\frac{2}{3}$
(4) $1\frac{2}{3}, 1\frac{1}{2}, 1\frac{5}{8}$
(5) $1\frac{5}{8}, 1\frac{2}{3}, 1\frac{1}{2}$

20. A school newsletter reported that 480 out of 640 students ride a bus to school. In lowest terms, what is the ratio of students who ride the bus to the total number of students?

Mark your answer in the circles in the grid on your answer sheet.

21. The instructions on a box of plant food call for 3 teaspoons of plant food for every 2 quarts of water. A bucket holds 16 quarts of water. Which of the following expressions could be used to find the number (n) of teaspoons of plant food you should add to the bucket of water?

(1) $\frac{3 \times 16}{2}$

(2) $\frac{2 \times 3}{16}$

(3) $\frac{2 \times 16}{3}$

(4) $\frac{16}{3 \times 2}$

(5) $3 \times 2 \times 16$

Question 22 is based on the following graph.

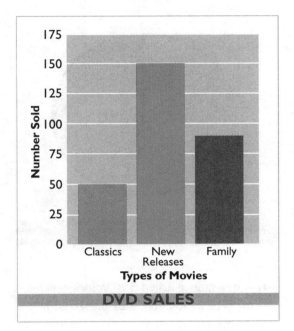

DVD SALES

22. Which of the following expressions best represents the ratio of the sales of new releases to the sales of classic DVDs?

(1) 1:4

(2) 1:3

(3) 1:2

(4) 2:1

(5) 3:1

Question 23 is based on the following graph.

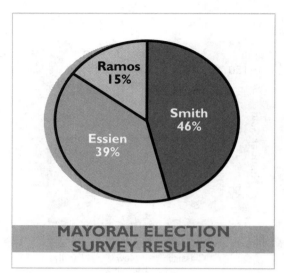

MAYORAL ELECTION SURVEY RESULTS

23. A newspaper asked 500 voters which of three candidates they preferred for mayor. Using the data, the newspaper created the graph shown above. What percent of the voters do NOT plan to vote for Essien?
    (1) 15%
    (2) 46%
    (3) 54%
    (4) 61%
    (5) 85%

24. A dowel is 235 centimeters long. Which of the following expressions could be used to convert the length to meters?
    (1) 235 × 10
    (2) 235 × 100
    (3) 235 ÷ 100
    (4) 235 × 1,000
    (5) 235 ÷ 1,000

25. How many fluid ounces are there in 2 quarts? (Use the facts 8 fluid ounces equal 1 cup, and 4 cups equal 1 quart.)
    (1) 16
    (2) 20
    (3) 32
    (4) 64
    (5) 128

Answers and explanations begin on page 17.

Math Pretest Answer Key

Part I

1. **(2) $61** With the coupon, the price at PC Central is $469 − $50 = $419. Subtract from $480 to find the difference in the prices: $480 − $419 = $61.

2. **(3) 12:30 P.M.** 240 ÷ 60 = 4. Start at 8:30 and add 4 hours. Tom will finish at 12:30 P.M.

3. **(5) $2,016** Add: 196 + 28 = 224 persons. Multiply by the cost per person: 224 × $9 = $2,016.

4. **(1) $7\frac{1}{8}$** Find the information in the table. Using fabric with a 45-inch width, you need $2\frac{3}{8}$ yards for one pair of pants. Multiply by 3: $2\frac{3}{8} \times 3 = \frac{19}{8} \times \frac{3}{1} = \frac{57}{8} = 7\frac{1}{8}$ yards.

5. **(5) Not enough information is given.** The question doesn't tell you whether the fabric on the bolt has a 45-inch or 60-inch width.

6. **17** Multiply: 6.8 × 2.5 = 17 gallons.

			1	7

7. **(2) $102** Find the cost of the copies at the regular rate: 3,000 × $0.04 = $120. Find the discount: $\frac{x}{\$120} = \frac{15}{100}$; $120 × 15 ÷ 100 = $18. Subtract: $120 − $18 = $102.

8. **(4) 2:5** Add to find the total number of classes: 16 + 24 = 40. Write the ratio of seniors to the total, and simplify: $\frac{16}{40} = \frac{2}{5}$.

9. **(2) $180** Find the first sale price: 20% of $250 is $50. Subtract: $250 − $50 = $200. Find the second sale price: 10% of $200 is $20. Subtract: $200 − $20 = $180.

10. **10125** The mean is the average. Add: 10,000 + 8,500 + 12,500 + 9,500 = 40,500. Divide by 4, the number of scores: 40,500 ÷ 4 = 10,125. Do not write the comma on the grid.

1	0	1	2	5

11. **(4) 80** There were 45 tuna and 35 cheese sandwiches sold. Add: 45 + 35 = 80.

12. **(1) 300** Linda sold about 25 ham sandwiches in one week. Multiply: 25 × 12 = 300.

13. **(4) 8.0** You need to find the perimeter, or distance, around the painting: 22 + 22 + 26 + 26 = 96 inches. To change to feet, use the fact 12 inches = 1 foot. Divide: 96 ÷ 12 = 8 feet.

Part II

14. **(5) 50,000** Look at the ten thousands place: 4̲8,164. The number to the right, 8, is greater than 5. Add 1 to the ten thousands place, and change the rest of the digits to zeros.

15. **(4) between $1,400 and $1,600** Use rounding to estimate an answer. Paul and Sandy will pay about $60 per month for about 20 months. $60 × 20 = $1,200. The down payment was about $300. Add: $1,200 + $300 = $1,500. Find the estimate in the range of answers.

16. 1080 From Rio Vista, the trip to Aldora and back is 90 × 2 = 180 miles. From Rio Vista, the trip to Sherborn and back is 120 × 2 = 240 miles. Multiply the distance to and from Aldora by 2 and the distance to and from Sherborn by 3. Add to find the total distance: (180 × 2) + (240 × 3) = 360 + 720 = 1,080 miles.

17. (1) 7.7 Add the amounts for Freeport and Boston: 5.3 + 9.4 = 14.7 million tons. Subtract from the amount for Long Beach: 22.4 − 14.7 = 7.7 million tons.

18. (2) 3 Divide: 28.2 ÷ 9.4 = 3. Without a calculator, the best way to solve the problem is to try each answer using estimation. 9 × 3 = 27, which is close to 28.2. The other choices lead to numbers much lower or higher than 28.2.

19. (3) $1\frac{1}{2}, 1\frac{5}{8}, 1\frac{2}{3}$ Each package weighs 1 pound and some fraction of a pound. Compare the fractions to arrange the weights in order. One way is to compare like fractions.

$1\frac{1}{2} = 1\frac{12}{24}$ $1\frac{5}{8} = 1\frac{15}{24}$ $1\frac{2}{3} = 1\frac{16}{24}$

$1\frac{12}{24} < 1\frac{15}{24} < 1\frac{16}{24}$, so $1\frac{1}{2} < 1\frac{5}{8} < 1\frac{2}{3}$

20. $\frac{3}{4}$ or .75 Write the ratio and simplify.

$\frac{480}{640} = \frac{3}{4} = .75$

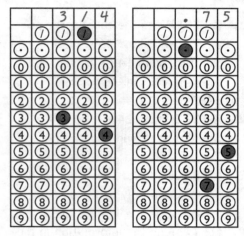

21. (1) $\frac{3 \times 16}{2}$ Set up the proportion: $\frac{3\text{ teaspoons}}{2\text{ quarts}} = \frac{x\text{ teaspoons}}{16\text{ quarts}}$. To solve the proportion, you would multiply 3 × 16 and divide by 2. Only option (1) shows this series of operations.

22. (5) 3:1 Compare the bars for new releases and classics. The bar for new releases (150) is three times larger than the bar for classics (50). You don't need to find the actual values for the bars; simply compare the sizes.

23. (4) 61% Subtract: 100% − 39% = 61%, or add: 15% + 46% = 61%.

24. (3) 235 ÷ 100 There are 100 centimeters in 1 meter. Divide by 100.

25. (4) 64 1 quart = 4 cups, so 4 × 8 fluid ounces = 32 fluid ounces in a quart. Find the fluid ounces in 2 quarts: 32 × 2 = 64 fluid ounces.

Evaluation Chart for Math Pretest

This evaluation chart will help you find the strengths and weaknesses in your understanding of math.

Follow these steps:
- Check your answers using the Answers and Explanations on pages 17–18.
- Circle the questions you answered correctly.
- Total your correct answers for each row.
- Use the results to see which math areas you need to study. If you got more than two wrong in any section, review the programs and workbook pages listed.

Questions	Total Correct	Program
1, 2, 3, 14, 15, 16	_____ / 6	20: Number Sense (pp. 20–39) 21: Problem Solving (pp. 40–59)
6, 17, 18	_____ / 3	22: Decimals (pp. 60–79)
4, 5, 19	_____ / 3	23: Fractions (pp. 80–99)
7, 8, 9, 20, 21, 22	_____ / 6	24: Ratio, Proportion, and Percent (pp. 100–119)
13, 24, 25	_____ / 3	25: Measurement (pp. 120–139)
10, 11, 12, 23	_____ / 4	26: Data Analysis (pp. 140–159)
TOTAL	_____ / 25	

Number Sense

1. Think About the Topic

The program that you are going to watch is about *Number Sense*. The video will show you different ways to think about using math. You will see people who use math in their work and who teach math. They will show that you use math all of the time without realizing it. In the course of daily life, you develop your own ways of solving problems with numbers. In other words, you already have some number sense.

2. Prepare to Watch the Video

Before you watch the program, use your number sense to answer these questions.

What is the quickest way to solve 250 + 150 in your head?

_____

You may have thought, *First I'll add 50 + 50 = 100. Then I'll add the 100 to 200 + 100 to get 400.*

However you solved the problem in your head, you used your number sense.

If you have to get to a place that you have never been to before, what do you do?

_____

You may have written something like, *I ask someone for directions and draw myself a map.* This question is related to using strategies to solve problems—something you will have to develop to pass the GED Math Test.

LESSON GOALS

MATH SKILLS

- Understand place value
- Compare and order numbers
- Use number lines and number patterns

GED PROBLEM SOLVING

- Understand math problems

GED MATH CONNECTION

- Explore calculator basics

GED REVIEW

EXTRA PRACTICE PP. 160–163

- Place Value
- Comparing and Ordering Numbers
- Number Patterns
- Calculator Basics

3. Preview the Questions

Read the questions under *Think About the Program* below, and keep them in mind as you watch the program. You will be reviewing them after you watch.

4. Study the Vocabulary

Review the terms to the right. Understanding the meaning of math vocabulary will help you understand the video and the rest of this lesson.

WATCH THE PROGRAM

As you watch the program, pay special attention to the host who introduces or summarizes major ideas that you need to learn about. The host may also tell you important information about the GED Math Test.

AFTER YOU WATCH

I. Think About the Program

How does comparing numbers help you in everyday life?

Why do some people call mathematics a "language?"

What is an example of showing an amount in different forms (fraction, decimal, and percent)?

In what types of situations is it helpful to estimate?

2. Make the Connection

The program talks about planning ahead when it comes to spending money. Do you have a goal of buying something? What is it? Do you have a plan or budget? If not, how could you start to make one?

digits—the symbols (0, 1, 2, 3, 4, 5, 6, 7, 8, 9) used to represent numbers

even numbers—numbers that can be evenly divided by 2; even numbers end in 2, 4, 6, 8, or 0

multiples—the result of multiplying one number by another number; for example, multiples of 4 are 4, 8, 12, 16, 20, and so on

number pattern—a series of numbers with a repeating relationship between each number and the next one in the series

number sense—common sense about numbers

odd numbers—numbers that cannot be evenly divided by 2; odd numbers end in 1, 3, 5, 7, or 9

place value—the value given to a digit based on its place in a number

whole numbers—counting numbers such as 1, 2, 3, 4, 5, 6, 7, 8, 9, 10, 11, and so on

> *"We all use numbers every day. How much do groceries cost? Which side of the street is that address on?"*

Number Sense

How Do You Use Math?

Math is a tool you use every day. It helps you figure out whether you have enough money with you to pay for your groceries or how much time to leave yourself to run errands.

Most everyday math doesn't require paper and pencil or a calculator. Most of the time, you solve math situations using your **number sense.** Number sense is common sense about numbers.

EXAMPLE: Suppose you and a friend go to a diner. Together you have $12. You both want a sandwich, drink, and fries. How could you figure out whether you have enough money?

You could round each item to the nearest dollar and add all the amounts mentally. You could also add to find the cost of one meal and then multiply the sum by 2. Either method will find the total cost of the bill. Then you could compare the total to $12.

As you can see, there are often several approaches to a problem. You may have thought of another method that would work just as well.

Trust your own common sense about numbers. Build on what you already know. As you practice your basic math skills, your number sense will improve, and your confidence in your math abilities will increase.

How Do You Feel About Learning Math?

Improving your math skills may help you get a better job or complete your education. You are probably looking forward to achieving these goals. Even so, you may be feeling a little nervous about the work that lies ahead.

Your feelings and attitudes about math will be an important factor to your success in learning math. Most people have a little math anxiety. Some believe they just don't have what it takes to learn math. That isn't true. You can learn math if you apply good study habits and keep practicing. The tips on the next page will help you succeed.

Tips for Success

- **Set aside some time each day to learn math.** Choose a time when your mind is rested and you feel awake.
- **Practice your math skills every day.** Learning math is like learning to play a musical instrument. Don't try to do a month's worth of work in one day. Take it step by step and keep practicing.
- **Always relate what you are learning to your life.** Try to apply your new skills to your everyday activities.
- **When you have difficulty with a problem, try a new approach.** Perhaps it will help to draw a picture or act out the situation.

NUMBER SENSE ▪ PRACTICE 1

Answer each question about your personal math and problem-solving skills.

1. Think about the past week. Choose three experiences, and explain how you used math to make decisions or solve problems.

2. You have a job interview in an unfamiliar part of town. What do you do to make sure you can find the location on time? *(Choose one or write one of your own.)*
   _____ **a.** Ask a friend to draw a map for you.
   _____ **b.** Find the address on a city map.
   _____ **c.** Ask a friend to help you make a list of directions.
   _____ **d.** Other: _____

3. Suppose you want to buy a new TV set that costs $300. You can finance the TV set and pay $39 per month for 10 months. What would you do to decide whether or not you should make this purchase? *(Choose one or write one of your own.)*
   _____ **a.** Add up this month's bills to find if there is $39 left over.
   _____ **b.** Find the total cost of the TV set if you finance it.
   _____ **c.** Call other stores to see what deals they offer for the TV set.
   _____ **d.** Other: _____

4. You go to the pharmacy to buy cough medicine. The smaller bottle costs about $6. The larger bottle costs a few dollars more, but it holds almost twice as much. What would you do to decide which bottle to buy? *(Choose one or write one of your own.)*
   _____ **a.** Read the label on the shelf to see which size is the best buy.
   _____ **b.** Think about how much medicine you will actually use and choose the size that fits your needs.
   _____ **c.** Ask the pharmacist to explain the difference in the prices.
   _____ **d.** Other: _____

Answers and explanations start on page 200.

How Do You Learn to Use Math?

While a math problem may have only one right answer, there may be more than one way to find that answer. There are many strategies for solving problems, and there are many ways to learn. How would you solve the example below?

E X A M P L E : Colin is standing in line to buy a movie ticket. There are twice as many people standing behind Colin as there are in front of him. If there are ten people in the line, how many people are standing in front of Colin?

Method 1: Use logical thinking. There are 10 people in the line. Subtract Colin and there are 9 left. Now think of all the number pairs that add to 9. The possibilities are:

    1 and 8   2 and 7   3 and 6   4 and 5

The problem tells you that there are twice as many people behind Colin as in front of him. Only the number pair 3 and 6 has one number that is twice the size of the other number.

Answer: There are three people in front of Colin.

Method 2: Draw a diagram. You know there are 10 people in the line, so draw 10 boxes.

Then experiment. Working from the left, use your finger to cover boxes one at a time. Your finger represents Colin's position in line. Keep moving to the right. Stop when you have twice as many boxes to the right of your finger as you do to the left.

You can see from the diagram that there must be 3 people in front of Colin. Think about your approach to this problem.

- If your approach was similar to Method 1, you like working with words and symbols.

- If your approach was more like the one used in Method 2, you like to see what is happening in a problem. Drawings, charts, and graphs can help you understand math.

There are also other methods that could be used to solve this problem. You could act it out with a group of nine other people or use objects to represent the people in line. Understanding how you learn best is an important part of succeeding in math.

Sometimes you may see another way to solve a problem. Try it your way. If you get the same answer, your way is probably best for you. Remember, building on your own number sense is the best way to learn math.

NUMBER SENSE ■ PRACTICE 2

Read each problem, then decide what strategy you would use to solve it. Choose from the list, or write a strategy of your own.

1. You want to buy a new sofa for your living room. What would you do to be sure the sofa in the store will fit along the wall in your house?

   _____ **a.** Make a scale drawing of the room and take it to the store with you.

   _____ **b.** Estimate the length of the wall by walking heel to toe. Then use the same method to measure the sofa in the store.

   _____ **c.** Use a tape measure to find the exact length of the wall. Then measure the sofa in the store with the tape measure.

   _____ **d.** Other: _____

2. You need to run an errand and finish your math homework by 8 P.M. You estimate that the errand will take 45 minutes and the math will take 1 hour. How do you decide the latest time you can leave on the errand and finish everything by 8 P.M.?

   _____ **a.** Add the times and subtract from 8 P.M.

   _____ **b.** Since 45 minutes is almost an hour, estimate that it will take 2 hours to do everything. Subtract 2 hours from 8 P.M.

   _____ **c.** Look at a clock and count backward from 8 P.M.

   _____ **d.** Other: _____

3. You know that Lindon is 50 miles east of Lake City and Rockford is 30 miles west of Lindon. Is Rockford east or west of Lake City? How would you figure it out?

   _____ **a.** Use a calculator.

   _____ **b.** Draw a map of the situation.

   _____ **c.** Talk it out with a friend.

   _____ **d.** Other: _____

4. You just started a job and your work offers three different health plans. How would you decide which health plan would be best for you and your family?

   _____ **a.** Make a chart of the good things and bad things about each plan.

   _____ **b.** Estimate your average health costs for a year and then use a calculator to see which of the plans would save the most money.

   _____ **c.** Ask the other workers in the office which plan they chose, and choose the most popular plan.

   _____ **d.** Other: _____

Answers and explanations start on page 200.

"Numbers were invented to count, compare, measure, and combine things. They represent things that matter to us."

Number Relationships

Place Value

Our number system uses only ten **digits:** 0, 1, 2, 3, 4, 5, 6, 7, 8, and 9. Using these digits, you can write any number.

You already use place value to understand the size of numbers. For instance, would you rather win $150 or $1,500? The difference in the numbers depends on **place value.** In other words, the value of a digit depends on its place in the number. The digit 1 in $1,500 means "one thousand," while the digit 1 in $150 means only "one hundred."

The diagram below shows the value of the first ten number places.

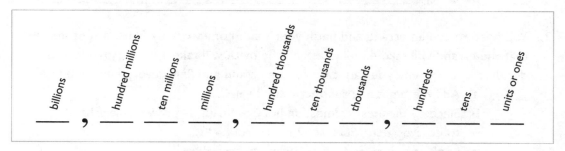

E X A M P L E : What is the value of the digits 2, 5, 7, and 8 in the number 250,078? Starting on the right, line up the number with the columns on the place-value chart.

Answer: The 2 is in the *hundred thousands* place; its value is $2 \times 100,000 = 200,000$.
The 5 is in the *ten thousands* place; its value is $5 \times 10,000 = 50,000$.
The 7 is in the *tens* place; its value is $7 \times 10 = 70$.
The 8 is in the *ones* place; its value is $8 \times 1 = 8$.

Note: While you don't read the 0s in the problem, they are needed to "hold" the values of the *thousands* and *hundreds* places. The number 250,078 is not the same as 2,578.

Reading and Writing Numbers

Your understanding of place value will help you read and write numbers. Numbers are read from left to right in groups of up to three digits. When you come to a comma, you also read the special name *(million, thousand,* or *hundred)* for that group of digits.

E X A M P L E : Write the number 5,604,070 in words. As you read the number, write the words. Put commas in the same place as their position in the number.

Answer: Five million, six hundred four thousand, seventy

EXAMPLE: Write the number *sixteen million, five thousand, three hundred forty-three* using digits.

Put the digits in the correct place-value columns. Use zeros as placeholders whenever a column does not have any value.

Answer: 16,005,343

Notice that the word *and* is not used when reading or writing numbers. The word *and* will have a special use when you begin working with decimal numbers.

NUMBER SENSE ▪ PRACTICE 3

A. Write the value of each underlined digit.

Examples: 4_2_0 $4 \times 100 = 400$ 4_2_0 $2 \times 10 = 20$

1. 97_2_ _____
2. _5_00 _____
3. 1,6_4_7 _____
4. _3_,004 _____
5. 5,_8_69 _____

6. _7_6,982 _____
7. 10_4_,927 _____
8. _3_57,641 _____
9. _5_,411,000 _____
10. 2,095,_6_23 _____

B. Write each number in words.

Example: 1,023   *one thousand, twenty-three*

11. 558 _____
12. 2,803 _____
13. 12,560 _____

14. 300,692 _____
15. 2,507,000 _____
16. 5,008,950 _____

C. Write each number by using digits.

Example: forty thousand, five hundred   _40,500_

17. three hundred thirty-six _____
18. four thousand, one hundred seven _____
19. twenty-four thousand _____
20. one hundred seventy thousand _____
21. fifteen thousand, ninety-eight _____
22. six million, forty thousand, nine _____

Answers and explanations start on page 200.
For more practice with place values, see page 160.

Comparing Numbers

Your understanding of place value helps you know whether one number is greater than or less than another number.

EXAMPLE: Mitch needs to earn some extra money. His uncle will pay him $82 to work in his store for a day, or he could work for a friend as a housepainter for $108 per day. He will take the job that pays more. Which job will he take?

Answer: Mitch will take the **house-painting job,** which pays $108 per day and is certainly more than $82.

Comparing numbers is one way to understand more about the relationship between the numbers. In math, these relationships can be written using the symbols <, >, and =.

Symbol		Example
< means *is less than*	4 < 9	4 *is less than* 9
> means *is greater than*	6 > 5	6 *is greater than* 5
= means *equals* or *is equal to*	7 = 7	7 *is equal to* 7

Hint: When using the symbols < and >, the small point on the symbol always points to the smaller number.

Use this rule to compare whole numbers:
- If both whole numbers have the same number of digits, work from left to right and compare the place-value columns.
- When you come to a place value with different digits, compare the digits. The greater digit is in the greater number.

EXAMPLE: Which is a true statement, 53 < 110 or 53 > 110?

Answer: 53 < 110 This means, "53 is less than 110." The symbol is pointing to the smaller number.

EXAMPLE: Which is greater, 3,571 or 3,538?

Compare each place-value column, starting from the left.
The digits are the same in the thousands and hundreds columns.
The digits are different in the tens column: 3,5**7**1   3,5**3**8
Compare 7 and 3. Since 7 is greater than 3, then 3,571 > 3,538.

Answer: The number **3,571** is greater.

Ordering Numbers

By comparing values, you can arrange numbers in order. Sometimes you will be asked to **order numbers** from greatest (largest or highest) to least (smallest or lowest). When comparing more than two numbers, write them in a column, lining up the ones column.

EXAMPLE: Write the numbers 156, 1,389, and 1,064 in order from <u>least (smallest)</u> to <u>greatest (largest)</u>.

Write the numbers in a column:

156
1,389
1,064

The number 156 is the smallest because it has the fewest digits. Now compare 1,389 and 1,064, starting at the left: 1,064 < 1,389 because in the hundreds column, 0 < 3.

Answer: From least to greatest, the order is: **156, 1,064, 1,389.**

NUMBER SENSE ▪ PRACTICE 4

A. Compare each pair of numbers. Write <, >, or = in the blank.

1.	590 _____ 490	7.	74,095 _____ 73,995
2.	3,068 _____ 3,608	8.	101,500 _____ 101,500
3.	184 _____ 184	9.	327 _____ 347
4.	305,060 _____ 304,060	10.	25,500 _____ 24,900
5.	1,200,000 _____ 2,100,000	11.	4,963,450 _____ 4,963,550
6.	1,004 _____ 1,040	12.	7,604 _____ 7,406

B. Solve each problem by comparing numbers.

13. For Saturday's game, the Dodgers sold 32,195 tickets. The Angels sold 32,300. Which team sold fewer tickets?

<u>Questions 14 and 15</u> refer to the table below.

Company	Donation
Food Saver	$15,700
Greenscape	$9,450
Value-Save Stores	$10,540
Barker Realty	$15,070

14. Which company donated the greatest amount of money?

15. Which company gave the smallest donation?

16. Write the names of the cities below in order by population from LEAST TO GREATEST.

City	Population
Baker	216,000
Clayton	119,300
Greenview	206,900
Maywood	117,850

17. The monthly salaries for three employees are $2,158, $1,872, and $2,827. Write the salaries in order from GREATEST TO LEAST.

18. AB Manufacturing earned profits of $4,355,900 in January. RS Industries earned $4,360,000 in the same month. Which company earned more in January?

Answers and explanations start on page 200.
For more practice comparing and ordering numbers, see page 161.

MATH SKILLS

"Mathematics is a way to describe the world. Learning the words, symbols, and rules for how things are done is part of learning the language."

Number Patterns

The Number Line

Whole numbers start at 0 and include all the numbers we use for counting. One way to show this idea is with a number line. Each mark represents a whole number. The arrow on the right end of the line shows that whole numbers continue without stopping.

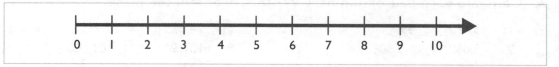

Finding Patterns

A **number pattern** is a series of numbers. For a pattern to exist, there must be some rule that describes the relationship between each number and the next one in the series. Take a look at the number pattern below. How would you describe it?

2   4   6   8   10. . .

There are many ways to describe this pattern.

- You may recognize it as a list of **even numbers.**
- You could also say it is a list of the numbers that can be evenly **divided by 2.**
- You may have noticed that 2 is added to each number to find the next number in the sequence. We say the rule for the pattern is "**add 2**" or "**counting by 2s.**"
- You could call it a list of the **multiples** of 2. The **multiples** of a number are the result of multiplying the number by 1, then 2, then 3, and so on.

All of these statements are different ways of saying the same thing, and all could be used to find the next number in the series, 12.

The next pattern may seem more difficult. To find the rule behind the pattern, pay close attention to the amount the numbers increase each time.

E X A M P L E : What number comes next in the pattern below?

1   2   4   7   11. . .

Answer: The pattern adds 1, then 2, then 3, and so on. The next number is **16.**

A. For each pattern, find the number that comes next.

1. 1   3   5   7   9   ____

2. 17   14   11   8   5   ____

3. 40   50   60   70   80   ____

4. 16   24   32   40   48   ____

5. 30   26   22   18   14   ____

6. 2   3   5   8   12   ____

7. 18   24   30   36   42   48   ____

8. 150   250   350   450   550   ____

B. Answer each question.

9. How many blocks will you need to build the next figure in this sequence?

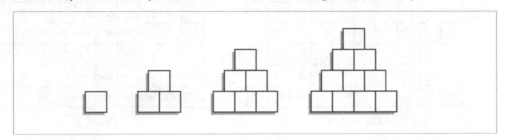

10. Draw the symbol that comes next in this pattern.

C. Solve.

11. In January, Lynn had $15 in her savings account. In February, she had $25. She had $35 in March and $45 in April. If the pattern continues, how much will she have in her savings account in May?

12. Daniel is training for a cross-country race. For the first week of training, he runs 4 laps every day. For the second week, he runs 8 laps per day. For the third week, he increases to 12 laps daily. For the fourth week, he runs 16 laps daily. If he continues the pattern, how many laps will he run each day during the fifth week?

13. Jane programmed her phone to dial a number every few minutes. The phone dials the number at 1:20 P.M., 1:40 P.M., 2:00 P.M., and 2:20 P.M. If the pattern continues, at what time will the phone dial the number again?

14. What letter comes next in this series?
M   T   W   T   F   S   ____

Hint: This pattern is a little tricky. It doesn't involve numbers. Think of something from everyday life.

Answers and explanations start on page 200.
For more practice with patterns, see page 162.

MATH SKILLS

Seeing Number Patterns

Numbers can be easier to understand when they are organized in a chart. Chart 1 organizes the first 100 numbers in rows of 10. In Chart 2, some of the multiples of 6 (6 × 1, 6 × 2, 6 × 3, and so on) are shaded.

1	2	3	4	5	6	7	8	9	10
11	12	13	14	15	16	17	18	19	20
21	22	23	24	25	26	27	28	29	30
31	32	33	34	35	36	37	38	39	40
41	42	43	44	45	46	47	48	49	50
51	52	53	54	55	56	57	58	59	60
61	62	63	64	65	66	67	68	69	70
71	72	73	74	75	76	77	78	79	80
81	82	83	84	85	86	87	88	89	90
91	92	93	94	95	96	97	98	99	100

Chart 1

1	2	3	4	5	6	7	8	9	10
11	12	13	14	15	16	17	18	19	20
21	22	23	24	25	26	27	28	29	30
31	32	33	34	35	36	37	38	39	40
41	42	43	44	45	46	47	48	49	50
51	52	53	54	55	56	57	58	59	60
61	62	63	64	65	66	67	68	69	70
71	72	73	74	75	76	77	78	79	80
81	82	83	84	85	86	87	88	89	90
91	92	93	94	95	96	97	98	99	100

Chart 2

Take a moment to complete Chart 2 by shading the remaining multiples of 6. Now study the pattern you have created. What do you notice about multiples of 6?

You may have noticed these properties:

- All the multiples of 6 are even numbers. None are odd numbers.
- The shaded boxes make a diagonal (slanted) pattern.
- The ones digits of the multiples form a repeating pattern: 6   12   18   24   30   36   42   48   54   60

If you study patterns of multiples, you will have an easier time recalling your multiplication and division facts. To learn how different patterns of multiples are related, shade in the multiples of two different numbers on the same chart.

EXAMPLE: The chart below shows the multiples of 10 shaded. Using another color, shade in the multiples of 5. How are the multiples of 5 and 10 related?

1	2	3	4	5	6	7	8	9	10
11	12	13	14	15	16	17	18	19	20
21	22	23	24	25	26	27	28	29	30
31	32	33	34	35	36	37	38	39	40
41	42	43	44	45	46	47	48	49	50
51	52	53	54	55	56	57	58	59	60
61	62	63	64	65	66	67	68	69	70
71	72	73	74	75	76	77	78	79	80
81	82	83	84	85	86	87	88	89	90
91	92	93	94	95	96	97	98	99	100

Answer: All the multiples of 10 are also multiples of 5.

Other important conclusions:
- All the multiples of 10 have a 0 in the ones place.
- The multiples of 5 end in either 0 or 5.

A. Shade in the multiples as directed, and answer the questions.

1. Shade in the multiples of 9.

1	2	3	4	5	6	7	8	9	10
11	12	13	14	15	16	17	18	19	20
21	22	23	24	25	26	27	28	29	30
31	32	33	34	35	36	37	38	39	40
41	42	43	44	45	46	47	48	49	50
51	52	53	54	55	56	57	58	59	60
61	62	63	64	65	66	67	68	69	70
71	72	73	74	75	76	77	78	79	80
81	82	83	84	85	86	87	88	89	90
91	92	93	94	95	96	97	98	99	100

2. How would you describe the pattern formed by the multiples of 9?

3. If the chart continued, what would be the next three multiples of 9?

4. Shade in the multiples of 4 and 8. Use two different colors.

1	2	3	4	5	6	7	8	9	10
11	12	13	14	15	16	17	18	19	20
21	22	23	24	25	26	27	28	29	30
31	32	33	34	35	36	37	38	39	40
41	42	43	44	45	46	47	48	49	50
51	52	53	54	55	56	57	58	59	60
61	62	63	64	65	66	67	68	69	70
71	72	73	74	75	76	77	78	79	80
81	82	83	84	85	86	87	88	89	90
91	92	93	94	95	96	97	98	99	100

5. What relationship do you see between the multiples of 4 and 8?

6. Vanessa claims that 153 is a multiple of 8. How can you use your understanding of number patterns to prove she is wrong?

B. Fill in the blanks to complete each pattern. You may use the blank chart below.

7. 15   18   ___   24   27   30   ___

8. 7   14   21   ___   ___   42   49

9. 12   24   ___   48   ___   72   84

10. 77   88   99   ___   ___   132

1	2	3	4	5	6	7	8	9	10
11	12	13	14	15	16	17	18	19	20
21	22	23	24	25	26	27	28	29	30
31	32	33	34	35	36	37	38	39	40
41	42	43	44	45	46	47	48	49	50
51	52	53	54	55	56	57	58	59	60
61	62	63	64	65	66	67	68	69	70
71	72	73	74	75	76	77	78	79	80
81	82	83	84	85	86	87	88	89	90
91	92	93	94	95	96	97	98	99	100

C. Answer the questions.

11. Which of the following numbers are multiples of both 3 <u>and</u> 5?

    12   15   25   30   35   40   45

12. Alex says, "Since 8 is a multiple of 4, every number that ends with an 8 must be a multiple of 4." Is this statement true or false? Explain your reasoning.

Answers and explanations start on page 200.
For more practice with patterns, see page 162.

MATH SKILLS

Understand Math Problems

In the video program, many of the people solved math problems. Maybe the situations didn't seem like math to you, but activities such as buying a car and measuring ingredients require math problem-solving skills.

A **math problem** presents a situation—either in words or in words and pictures—that requires you to find a solution.

On the GED Math Test, problems will not be as straightforward as the following: $785 - 300 - 340$. On the GED, a situation will be presented, and you will have to decide how to use the information to solve the problem.

Here are two problems that you might see on the GED. What is the question asking you to find in each example?

EXAMPLE 1: Linda earned $189 in tips last week and $213 in tips this week. If she also earns $245 in wages each week, how much did Linda earn in wages for the 2 weeks?

Read the problem carefully and think about what the question is asking. In other words, what were Linda's <u>wages</u> for 2 weeks?

EXAMPLE 2: Linda earned $189 in tips last week and $213 in tips this week. She also earns $245 in wages each week. How much did Linda earn in tips and wages for the 2 weeks?

Think about this situation. The question asks you to find the total amount Linda earned in <u>tips and wages</u> for both weeks.

Multiple-Choice Questions

Most of the problems on the GED Math Test (40 out of 50 problems) will be in **multiple-choice** format. That means that there will be five answer choices to choose from. Only one answer choice is correct, but you have to be careful. The wrong answers are often based on common mistakes in problem solving and critical thinking, so read the problem carefully.

EXAMPLE 3: Linda earned $190 in tips last week and $210 in tips this week. If she earns $200 in tips next week, how much money will Linda earn in tips for the three weeks?

(1) $190   (incorrect; shows only one of the amounts to be added)
(2) $200   (incorrect; shows only one of the amounts to be added)
(3) $390   (incorrect; adds only two of the amounts)
(4) $410   (incorrect; adds only two of the amounts)
(5) $600   (correct; adds $190 plus $210 plus $200)

Standard Grid Questions

Another GED question format is the **standard grid.** Eight out of the 50 problems on the GED Math Test require you to solve a problem that does not have multiple-choice options.

Instead of choosing from five answer choices, you will need to fill in your answer on a grid similar to the one shown here. You will learn more about these grids later in this book.

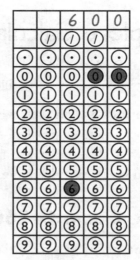

Answer to Example 3

GED MATH PRACTICE

UNDERSTANDING MATH PROBLEMS

Choose the <u>one best answer</u> to each question below. For questions 1 through 3, choose the option that best restates the question.

1. Alex walks about 3 miles a day, 5 days a week. If Alex can walk 1 mile in about 20 minutes, how many minutes would it take him to walk 3 miles?
   - **(1)** How many miles does Alex walk in one week?
   - **(2)** How long does it take Alex to walk three miles?

2. Carla earns $15,000 each year. If she works 50 weeks out of the year, how much does Carla earn each month?
   - **(1)** How much money does Carla make each month?
   - **(2)** How much money does Carla make each week?

3. Tanya travels 20 miles to work every day. If she works 5 days a week, how many miles does Tanya travel to and from work each week?
   - **(1)** How far does Tanya travel to and from work each day?
   - **(2)** How far does Tanya travel to and from work each week?

<u>Questions 4 and 5</u> refer to the chart below.

Trail Mix Shipment	
Tropical trail mix	25 pounds
Student trail mix	40 pounds
Basics trail mix	30 pounds
Sunshine trail mix	35 pounds

4. A store receives a shipment of different trail mixes. Which of the following lists the weights of the shipment in order from GREATEST TO LEAST?
   - **(1)** tropical, student, basics, sunshine
   - **(2)** student, basics, sunshine, tropical
   - **(3)** tropical, basics, sunshine, student
   - **(4)** student, sunshine, basics, tropical
   - **(5)** basics, tropical, sunshine, student

5. The shipment weights follow a pattern. If the pattern continues, what would be the next heaviest shipment weight, in pounds?
   - **(1)** 15   **(4)** 50
   - **(2)** 20   **(5)** 55
   - **(3)** 45

Answers and explanations start on page 201.

Explore Calculator Basics

The GED Math Test is divided into two parts. Each part has 25 questions, for a total of 50 questions. Only Part I allows calculators, which are supplied when you take the test.

Calculators on Part I make it possible to (1) solve problems with realistic numbers and (2) focus more on problem-solving skills. The calculator can be a helpful tool when used wisely. Not every problem needs a calculator. The important thing is to use it when it would save time or be more accurate.

The Casio *fx*-260 Calculator

The Casio *fx*-260 is the official calculator of the GED Math Test. A brief introduction to the Casio *fx*-260 is shown below. The basic operations will be explained later in this book. On the GED Math Test, you will need to use only a few of the calculator functions, so don't worry. This will be easier than it may seem at a glance!

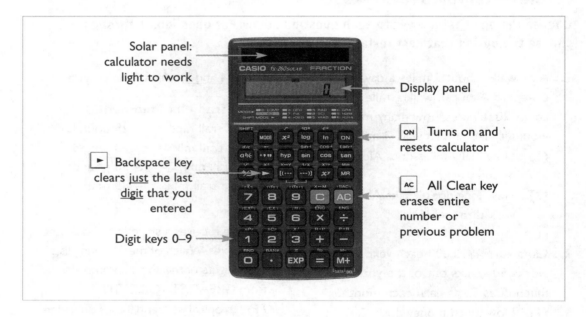

Enter the following examples on your calculator. Compare your display to those shown.

EXAMPLE 1: [AC] [2] [5]   [25.]   Always start with [AC] to make sure your calculator's memory is clear.

EXAMPLE 2: [AC] [3] [5] [0] [0]   [3500.]   On the calculator, commas are not used in numbers like 3,500.

EXAMPLE 3: [AC] [7] [2] [2] [►] [3]   [723.]   If you make a mistake, you can use the backspace key to erase just the 2 and replace it with a 3.

* If you have a different calculator, you may need to find how to clear entries.

Note that many calculators have a [CE] or [C] key to clear entries.

CALCULATOR BASICS

A. **Match each number below with the correct calculator display.**

Examples

For 8,480, the display shows | 8480. |

For 84,800, the display shows | 84800. |

_____ **1.** 545	**a.**	54000.
_____ **2.** 5,040	**b.**	5400.
_____ **3.** 54,000	**c.**	545.
_____ **4.** 5,400	**d.**	500.
_____ **5.** 500	**e.**	5040.

B. **Enter each number below in your calculator. Clear your display before each entry.**

Examples

562   `AC` `5` `6` `2`

20,460   `AC` `2` `0` `4` `6` `0`

  6.   489

  7.   6,256

  8.   10,573

  9.   115,987

10.   3,468,920

11.   23,546,989

C. **Use the Backspace key or the Clear key on your calculator to make the corrections described below.**

Example

Enter 764. Correct your entry to show 765.   `AC` `7` `6` `4` `▶` `5` | 765. | *or*

                                           `AC` `7` `6` `4` `C` `7` `6` `5` | 765. |

12. Enter 224. Correct your entry to show 223.

13. Enter 1,005. Correct your entry to show 1,015.

14. Enter 9,868. Correct your entry to show 9,968.

15. Enter 78,323. Correct your entry to show 783,123.

16. Enter 20,050. Correct your entry to show 20,500.

Answers and explanations start on page 201.

For more practice with calculators, see page 163.

GED MATH CONNECTION

GED Review: Number Sense

Choose the <u>one best answer</u> to each question below.

1. Which of the following amounts represents the figure shown in the display?

102056.

   (1) 1,020
   (2) 1,256
   (3) 10,205
   (4) 12,056
   (5) 102,056

2. Which of the following amounts is the same as two hundred twenty thousand, forty-six?

   (1) 2,246
   (2) 22,046
   (3) 22,460
   (4) 220,046
   (5) 224,600

3. What is the value of the underlined digit in the number below?

   1,5<u>6</u>7,892

   (1) 6
   (2) 60
   (3) 600
   (4) 6,000
   (5) 60,000

Questions 4 through 6 refer to the chart below.

Kelso's Furniture Store		
Year	Revenue	Expenses
1	$132,987	$78,902
2	$134,998	$81,646
3	$138,067	$81,976
4	$140,964	$80,245
5	$141,006	$89,433

4. What are the two greatest revenue amounts?
   (1) $134,998 and $132,987
   (2) $140,964 and $134,998
   (3) $141,006 and $140,964
   (4) $141,006 and $132,987
   (5) $141,006 and $138,067

5. For which two years were expenses approximately equal?
   (1) Year 4 and Year 5
   (2) Year 3 and Year 5
   (3) Year 2 and Year 4
   (4) Year 2 and Year 3
   (5) Year 1 and Year 2

6. Which of the following lists the expenses in order from LEAST TO GREATEST according to year?
   (1) Year 1, Year 2, Year 3, Year 4, Year 5
   (2) Year 1, Year 4, Year 2, Year 3, Year 5
   (3) Year 1, Year 4, Year 5, Year 2, Year 3
   (4) Year 5, Year 3, Year 2, Year 4, Year 1
   (5) Year 5, Year 4, Year 3, Year 2, Year 1

7. Marcie is packing a box and wants to put the heaviest packages on the bottom and the lightest packages on top. If the packages weigh 8 pounds, 4 pounds, 10 pounds, and 6 pounds, which order shows the package weights listed from HEAVIEST TO LIGHTEST?
   - **(1)** 4, 6, 8, 10
   - **(2)** 4, 8, 6, 10
   - **(3)** 8, 4, 6, 10
   - **(4)** 10, 8, 4, 6
   - **(5)** 10, 8, 6, 4

8. What is the value of the 7 in 175,261?
   - **(1)** 70
   - **(2)** 700
   - **(3)** 7,000
   - **(4)** 70,000
   - **(5)** 700,000

9. Which of the following amounts is the same as fourteen thousand, seventy?
   - **(1)** 1,470
   - **(2)** 10,470
   - **(3)** 14,070
   - **(4)** 14,700
   - **(5)** 147,000

10. Which of the following expressions is true?
    - **(1)** 6,300 = 6,030
    - **(2)** 6,300 > 6,030
    - **(3)** 6,300 < 6,030
    - **(4)** 6,300 = 6,003
    - **(5)** 6,300 < 6,003

11. Marcus sets his alarm to go off at 6:15 every morning. The snooze button on his alarm clock gives him another 5 minutes before the alarm goes off again. If Marcus hits the snooze button 4 times before getting out of bed, at what time does he get up?
    - **(1)** 6:35
    - **(2)** 6:30
    - **(3)** 6:25
    - **(4)** 6:20
    - **(5)** 6:15

12. What number comes next in the pattern below?

    48, 56, 64, 72, _____

    - **(1)** 74
    - **(2)** 76
    - **(3)** 78
    - **(4)** 80
    - **(5)** 88

13. Jean is trying to get in shape by swimming laps at the local pool. Each time she swims during Week 1, she swims 5 laps. She swims 8 laps during Week 2 and 11 laps during Week 3.
    If Jean continues at this rate, how many laps will she swim during Week 4?
    - **(1)** 17
    - **(2)** 14
    - **(3)** 11
    - **(4)** 8
    - **(5)** 5

Answers and explanations start on page 201.

Problem Solving

1. Think About the Topic

The program that you are going to watch is about *Problem Solving*. The video will describe strategies that you can use for approaching and solving different types of math problems.

2. Prepare to Watch the Video

As you watch the program, think about how you can use the strategies. For example, do you know when to estimate and when to find an exact answer?

A candidate wants to talk about the growing population in her district. Would she need an exact answer or an estimate?

_____

An *estimate* would give a good enough idea of the population.

At a wedding reception, how many people will be at the sit-down dinner? Do you need an estimate or an exact answer?

_____

An *exact answer* is needed so that there are enough dinners for everyone without the hosts paying for extra meals.

Deciding when you can estimate and when you need to find an exact answer is an important skill—in life and on the GED.

3. Preview the Questions

Read the questions under *Think About the Program* on the next page and keep them in mind as you watch the program. You will be reviewing them after you watch.

4. Study the Vocabulary

Review the terms to the right. Understanding the meaning of math vocabulary will help you understand the video and the rest of this lesson.

WATCH THE PROGRAM

As you watch the program, pay special attention to the host who introduces or summarizes major ideas that you need to learn about. The host may also tell you important information about the GED Math Test.

AFTER YOU WATCH

I. Think About the Program

What are the basic steps in the problem-solving process?

How would you approach a real-life problem that is different from any that you have seen before?

What are some of the different strategies you could use to help you understand a math problem?

The program talks about the order of operations. What is it? Why do you think a set order of operations is needed?

2. Make the Connection

The program talked about figuring interest and finding the real cost of an item, including the interest payment. Is there something that you would like to buy on credit? What information would you need to find the total cost, including interest? How would you get that information?

TERMS

compatible numbers— numbers that work well together

difference—the solution to a subtraction problem

estimation—finding an approximate amount

front-end estimation— using the first digits of numbers to estimate

partial products—the answers to the steps in a multi-step multiplication problem

product—the answer to a multiplication problem

quotient—the answer to a division problem

remainder—any leftover amount after finding the whole number part of the answer to a division problem

rounding—estimating to a certain place value

set-up problem— a multiple-choice problem with options that show *how* a problem could be solved

sum—the answer to an addition problem

"Doing well in math often means choosing the right operation."

Adding and Subtracting

Addition

Adding is combining like objects or numbers to find the total, or **sum**. Addition problems are written in rows or columns using the plus sign (+). Write the numbers in a column before you add, lining up like place-value columns.

E X A M P L E : Find the total of 243, 62, and 157.

Step 1. Write the numbers in a column. Line up the place-value columns.

Step 2. Work from <u>right to left</u>. Add the ones column. If the total of any column is 10 or more, **regroup,** or carry. When you regroup, you write the ones value of the column total under the column and write the tens value in the next column on the left.

Step 3. Add the tens and the hundreds columns.

$$\begin{array}{r} 243 \\ 62 \\ + 157 \\ \hline \end{array} \qquad \begin{array}{r} \overset{1}{243} \\ 62 \\ + 157 \\ \hline 2 \end{array} \qquad \begin{array}{r} \overset{11}{243} \\ 62 \\ + 157 \\ \hline 462 \end{array}$$

A special property of addition can make your work easier. You can add numbers in any order, and the answer will be the same.

Tip: When adding digits, look for digits that will combine to make ten.

E X A M P L E : Add: $4 + 8 + 3 + 2 + 6 + 5 + 7$.

There are three pairs of numbers that equal 10: 4 and 6, 8 and 2, and 7 and 3.
Rewrite the order of the numbers: $(4 + 6) + (8 + 2) + (3 + 7) + 5$
Add the 10s and the remaining digit: $10 + 10 + 10 + 5 = \mathbf{35}$

Mental Math

Mental math is solving a problem or even part of a problem in your head. If you compute math without pencil and paper or calculator, your math skills will improve. Take time to practice the mental math strategies you will find in this book. They can save you valuable time when you are taking a test.

To add mentally, start on the left to combine place values.

E X A M P L E : Add 47 and 35.
Think: Add the tens: $40 + 30 = 70$

Add the ones: $7 + 5 = 12$

Combine: $70 + 12 = \mathbf{82}$

A. Solve.

1.
```
    34
+   25
```

3.
```
   465
+   92
```

5.
```
  1,238
+ 4,057
```

7.
```
    16
    62
+  155
```

2.
```
   159
+   37
```

4.
```
  1,039
+ 2,634
```

6.
```
  3,054
     85
+   648
```

8.
```
   238
   446
+  572
```

B. Rewrite each problem in a column and solve.

9. 168 + 84 + 302

10. 1,073 + 748 + 1,947

11. 75 + 149 + 16

12. 350 + 928 + 415

C. Solve each problem using mental math.

13. 36 + 49

14. 57 + 85

15. 94 + 65

16. 18 + 96

17. 14 + 36

18. 76 + 53

D. Solve.

19. Maria's grades for four tests are shown below:

    Test 1: 78 points
    Test 2: 92 points
    Test 3: 86 points
    Test 4: 105 points

    How many points did she earn in all for the tests?

20. The Lighthouse Theater spent $1,568 to build the sets for a play. The costumes for the play cost $827. What was the total amount spent on sets and costumes?

Questions 21 and 22 refer to the following ad.

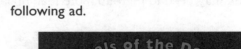

Deals of the Day!

Desktop Computer $1,199   Color Printer $159   Scanner $219

21. Karleen bought the desktop computer and the scanner shown in the ad. How much did she spend?

22. Craig bought a laptop computer for $1,385 and the printer in the advertisement. What was his total bill?

Answers and explanations start on page 201.
For more practice adding whole numbers, see page 164.

Subtraction

Subtraction is the opposite of addition. Addition combines amounts to find a sum. Subtraction takes away an amount to find what amount is left. The answer to a subtraction problem is called the **difference.** The minus sign (−) tells you to subtract.

Subtraction is used to compare numbers. When you compare numbers by subtracting, you are trying to find the difference between the numbers. In other words, you want to know how much more or less one number is than another.

To subtract, write the numbers so that the larger number is on top. Line up the place-value columns. Then start on the right in the ones place. You must regroup, or borrow, any time the digit you are subtracting is larger than the digit above it.

EXAMPLE: Subtract 58 from 182.

Step 1. Line up the place-value columns. You must write the smaller number on the bottom.

Step 2. Start in the ones column. Since 8 is greater than 2, you must regroup one 10 from the tens column.

Step 3. Subtract each column, working from right to left.

$$\begin{array}{r} 182 \\ -\ 58 \\ \hline \end{array} \qquad \begin{array}{r} {\scriptstyle 7\ 12} \\ 18\!\!\!/2 \\ -\ 58 \\ \hline \end{array} \qquad \begin{array}{r} {\scriptstyle 7\ 12} \\ 18\!\!\!/2 \\ -\ 58 \\ \hline 124 \end{array}$$

You can use addition to check the answer to a subtraction problem. Add your answer to the number you subtracted. The sum should be the number that you started with.

Check:
$$\begin{array}{r} {\scriptstyle 1} \\ 124 \\ +\ 58 \\ \hline 182 \end{array}$$

You may need to regroup more than once in a problem.

EXAMPLE: Subtract: 504 − 236

Step 1. Line up the place-value columns.

Step 2. You must regroup to subtract the ones column. Since the tens column has a zero, regroup ten 10s from the hundreds column. Then regroup one 10 from the tens column.

Step 3. Subtract each column, working from right to left.

$$\begin{array}{r} 504 \\ -\ 236 \\ \hline \end{array} \qquad \begin{array}{r} {\scriptstyle 4\ 10} \\ 50\!\!\!/4 \\ -\ 236 \\ \hline \end{array} \qquad \begin{array}{r} {\scriptstyle 9\ 14} \\ {\scriptstyle 4\ 10} \\ 50\!\!\!/4 \\ -\ 236 \\ \hline 268 \end{array}$$

Check: 268 + 236 = 504

Mental Math

Don't try to regroup when you subtract in your head. Instead, break the smaller number into parts, and subtract one part at a time.

EXAMPLE: Subtract: 85 − 47.

Think: I can break 47 into two parts: 40 and 7.
$$85 - 40 = 45 \text{ and } 45 - 7 = \textbf{38}$$

A. Solve.

1.	59 − 23	3.	62 − 46	5.	4,038 − 2,197	7.	924 − 179
2.	786 − 156	4.	348 − 175	6.	600 − 274	8.	5,382 − 3,275

B. Rewrite each problem in a column and solve. Check by using addition.

9. 562 − 89 =

10. 951 − 148 =

11. 90 − 38 =

12. 1,005 − 194 =

C. Solve each problem using mental math.

13. 94 − 12 =

14. 76 − 28 =

15. 106 − 45 =

16. 85 − 36 =

17. 57 − 15 =

18. 123 − 85 =

D. Solve.

Questions 19 and 20 refer to the following information.

> **Proposition C**
> **Election Results**
> Yes: 3,475
> No: 1,592

19. How many more people voted yes than voted no?

20. Only 2,315 people voted yes on Proposition D. How many more voted yes on Proposition C than on D?

21. David has driven 318 miles of a 600-mile trip. How many miles does he have left to drive?

22. Right now, Kira earns $1,926 per month. When she finishes school, she will qualify to work as a manager, which pays $2,620 per month. How much more per month will she earn if she gets the manager job?

23. Lamar has $2,564 in the bank. If he writes a check for $395, how much money will he have left in the bank?

Answers and explanations start on page 201.
For more practice subtracting whole numbers, see page 164.

"Take your time and write out each step. Scratch paper is cheap."

Multiplying and Dividing

Multiplication

Multiplying is a way of adding the same number many times. For instance, when you multiply 2 by 5, you are adding 2 five times: $2 + 2 + 2 + 2 + 2 = 10$ and $2 \times 5 = 10$. In the equation $2 \times 5 = 10$, the symbol $(\times)$ is the times sign, and the answer is called the **product.** You can say that the product of 2 and 5 is 10.

To multiply numbers with more than one digit, multiply each digit in the top number by each digit in the bottom number.

EXAMPLE: Find the product of 135 and 7.

Step 1. Write the number with more digits on top.

Step 2. Multiply each digit in the top number, starting in the ones column, by the bottom number. If the product of a column is 10 or more, regroup.

Step 3. After multiplying the next column, add the regrouped value.

$$
\begin{array}{r} 135 \\ \times\ 7 \\ \hline \end{array}
\qquad
\begin{array}{r} \overset{3}{1}35 \\ \times\ 7 \\ \hline 5 \end{array}
\qquad
\begin{array}{r} \overset{2\,3}{1}35 \\ \times\ 7 \\ \hline \mathbf{945} \end{array}
$$

When the number on the bottom has more than one digit, you must do the problem in stages. Working from right to left, multiply the top number by a digit in the bottom number. Then add these **partial products.** In the next example, the regrouping marks are not shown. As you gain experience, you may choose to regroup mentally.

EXAMPLE: Multiply: 368×124

Step 1. Write the number with more digits on top. Multiply the top number by 4. Write the partial product.

Step 2. Multiply the top number by 2. Since the 2 actually represents 2 tens, or 20, write a zero in the ones place of the partial product.

Step 3. Multiply the top number by 1. Since the 1 represents 100, write two zeros in the partial product. Finally, add the partial products.

$$
\begin{array}{r} 368 \\ \times\ 124 \\ \hline 1472 \end{array}
\qquad
\begin{array}{r} 368 \\ \times\ 124 \\ \hline 1472 \\ 7360 \end{array}
\qquad
\begin{array}{r} 368 \\ \times\ 124 \\ \hline 1\,472 \\ 7\,360 \\ 36\,800 \\ \hline \mathbf{45{,}632} \end{array}
$$

Working with Zeros

When a number ends in zero, you can use a shortcut to make your work easier. Look for a pattern in these examples.

$$
\begin{array}{r} 43 \\ \times\ 2 \\ \hline 86 \end{array}
\qquad
\begin{array}{r} 43 \\ \times\ 20 \\ \hline 860 \\ \text{one zero} \end{array}
\qquad
\begin{array}{r} 43 \\ \times\ 200 \\ \hline \mathbf{8{,}600} \\ \text{two zeros} \end{array}
$$

To multiply by a number ending in zero, write the total number of zeros in the product, and then multiply by the nonzero digit.

A. Solve.

1.
$$\begin{array}{r} 16 \\ \times\ 8 \\ \hline \end{array}$$

4.
$$\begin{array}{r} 168 \\ \times\ 7 \\ \hline \end{array}$$

7.
$$\begin{array}{r} 245 \\ \times\ 14 \\ \hline \end{array}$$

2.
$$\begin{array}{r} 25 \\ \times\ 9 \\ \hline \end{array}$$

5.
$$\begin{array}{r} 193 \\ \times\ 8 \\ \hline \end{array}$$

8.
$$\begin{array}{r} 708 \\ \times\ 29 \\ \hline \end{array}$$

3.
$$\begin{array}{r} 72 \\ \times\ 5 \\ \hline \end{array}$$

6.
$$\begin{array}{r} 724 \\ \times\ 5 \\ \hline \end{array}$$

9.
$$\begin{array}{r} 546 \\ \times\ 20 \\ \hline \end{array}$$

B. Rewrite each problem in a column and solve.

10. $123 \times 50 =$

11. $26 \times 54 =$

12. $208 \times 300 =$

13. $451 \times 60 =$

14. $913 \times 42 =$

15. $625 \times 200 =$

C. Solve.

16. Max pays $345 for rent each month. How much does he spend on rent in one year? (1 year = 12 months)

17. A computer can print 120 labels in 1 minute. How many labels can it print in 30 minutes?

18. Candice shot 5 rolls of film on her vacation. If each roll had 24 pictures, how many pictures did she take?

19. Jennifer's employer deducts $12 for health insurance from her weekly paycheck. After 42 weeks of work, how much has been deducted for health insurance?

Questions 20 and 21 refer to the table below.

CALORIES IN FOODS	
Food	**Calories**
Skim milk (1 cup)	85
Apple (medium)	80
Cheese pizza (1 slice)	145
Oat breakfast cereal (1 cup)	120

20. Nita drinks three cups of skim milk per day. How many calories does she get from skim milk in seven days?

21. Frank had cheese pizza for lunch. If he ate three slices, how many calories did he consume?

Answers and explanations start on page 202.
For more practice multiplying whole numbers, see page 165.

Division

Division is finding how many times one number can be subtracted from another number. Another way to look at division is to think how many times one number "goes into" another number. The answer to a division problem is called the **quotient.**

The long-division process is actually a shortcut. Instead of dividing a large number, we break it into smaller parts and divide only a few digits at a time.

EXAMPLE: Divide 1,416 by 6.

Step 1. Write the number to be divided inside the bracket. Since 6 is larger than 1, divide 6 into 14. Write 2 above the 4 in 14. Then multiply: 6 × 2 = 12. Subtract 12 from 14.

Step 2. Bring down the next digit, 1. Divide 6 into 21. Write 3 above the 1. Multiply 3 × 6 = 18, and subtract.

Step 3. Bring down the next digit, 6. Divide 6 into 36. Write the 6 above the 6. Multiply 6 × 6 = 36, and subtract. Since the result is 0, the division is finished and there is no **remainder.**

```
     2              23            236
 6)1,416        6)1,416        6)1,416
   12             12             12
    2             21             21
                  18             18
                   3             36
                                 36
                                  0
```

Division is the opposite of multiplication, so you can check division by multiplying. Multiply the quotient by the number you divided by. Add any remainder. The result should be the number you started with. *Check:* 236 × 6 = 1,416.

Dividing by more than one digit works the same way.

EXAMPLE: Divide 2,872 by 14.

Step 1. Since 14 is greater than 2, start with the first two digits. Divide 14 into 28. Write 2 above the 8 in 28. Multiply: 14 × 2 = 28, and subtract.

Step 2. Although the difference is 0, the division is not finished. Bring down the next digit, 7. How many times does 14 divide into 7? Write the 0 above the 7 and bring down the next digit, 2.

Step 3. Divide 14 into 72. The answer is 5. Multiply: 5 × 14 = 70, and subtract. There is a remainder of 2.

```
      2             20            205
 14)2,872      14)2,872      14)2,872
    28            28            28
     0            07            072
                                 70
                                  2
```

Answer: The quotient is **205 r2.** The letter *r* means remainder.

Check: Multiply: 205 × 14 = 2,870. Add the remainder: 2,870 + 2 = 2,872.

A. Solve. Check by using multiplication.

1. $5\overline{)810}$ 4. $13\overline{)3,679}$ 7. $4\overline{)615}$

2. $9\overline{)567}$ 5. $19\overline{)1,349}$ 8. $12\overline{)5,913}$

3. $7\overline{)2,212}$ 6. $28\overline{)14,112}$ 9. $32\overline{)19,424}$

B. Rewrite each problem using a division bracket, and solve. Check by using multiplication.

10. $5,056 \div 6$ 12. $1,701 \div 27$

11. $2,210 \div 23$ 13. $5,376 \div 34$

C. Solve.

14. The library committee raised $584 by selling tickets to a play. If the tickets were $4 each, how many were sold?

15. A bookstore paid $2,162 for an order of 94 copies of a new book. How much did the store pay for each book?

16. A factory assembly line can make 75 products per hour. Yesterday the assembly line completed 1,125 products. For how many hours did the assembly line run?

17. A sign at a car lot says:

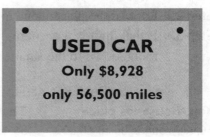

USED CAR
Only $8,928
only 56,500 miles

Jeff wants to pay for the car in 24 equal monthly payments. Before any interest is added, how much would Jeff owe each month?

18. Millie has to read a 960-page book in 6 weeks. How many pages will she have to read each week to finish on time?

Answers and explanations start on page 202.
For more practice dividing whole numbers, see page 165.

"A good process can help you break a problem into manageable steps."

Estimation

Rounding

In life we often use estimation to solve problems. **Estimation** means using approximate amounts instead of exact amounts to do calculations. **Rounding** is one way to estimate. When you round numbers, you make them simpler and easier to work with.

One way to round numbers is to use a number line.

EXAMPLE: Round 438 to the nearest hundred.

Is 438 closer to 400 or 500? As you can see, the number 450 is halfway

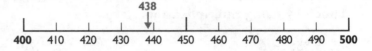

between 400 and 500. Any number less than 450 is closer to 400.

Answer: 438 rounds to **400**.

You can also round by using your understanding of place value.

Step 1. Choose the place value to which you want to round.
Step 2. Look at the place value to the right.
Step 3. If that digit is 5 or more, round up to the next larger number. If it is less than 5, round down to the smaller number.

EXAMPLES: Round 56,480 to the place values shown below.

Ten thousands place	Thousands place	Hundreds place
56,480 rounds up to 60,000	56,480 rounds down to 56,000	56,480 rounds up to 56,500

Estimation is an important tool in problem solving. When you start a new problem, first estimate an answer. Then solve the problem using the exact numbers. Once you have an answer, compare it to your estimate. The two numbers should be reasonably close. If they aren't close, rework the problem or check your work some other way.

EXAMPLE: Lois earns $85 per day at her job. She is scheduled to work 21 days in June. How much will she earn in June?

Step 1. Estimate an answer: $85 rounds up to $90, and 21 rounds down to 20 days. Multiply the rounded numbers: $90 × 20 = $1,800.
Step 2. Multiply the actual numbers: $85 × 21 = $1,785.
Step 3. Compare the actual answer to the estimate: $1,785 is close to $1,800.

Answer: Lois will earn **$1,785** in June.

A. Round each number as directed.

Example: Round 845 to the nearest hundred. _800_

1. Round 756 to the nearest ten. _____
2. Round 2,530 to the nearest thousand. _____
3. Round 19,615 to the nearest hundred. _____
4. Round 6,290,755 to the nearest million. _____
5. Round 14,935 to the nearest hundred. _____

B. Use rounding to estimate each answer. Then find the exact answer.

Example: 5,820 + 3,219   estimate: _9,000_   exact answer: _9,039_

6. 394 + 225 + 179   estimate: _____   exact answer: _____
7. 5,346 − 1,592   estimate: _____   exact answer: _____
8. 840 × 18   estimate: _____   exact answer: _____
9. 6,264 ÷ 12   estimate: _____   exact answer: _____

C. Use rounding to estimate each answer. Then find each exact answer.

10. Kenneth bought new dining room furniture for $1,032. He plans to pay for the furniture in 12 monthly payments. How much will he pay each month?

11. The land area of Minnesota is 79,617 square miles. The area covered by water is 7,326 square miles. What is the total area in square miles of Minnesota?

12. Bob ordered 280 Los Angeles Lakers caps to sell in his store. Each cap cost Bob $17. How much did Bob pay for the caps?

13. A school needs $28,000 to put in a new library. So far this year, the parents have raised $17,642. How much more do they need for the library?

14. An electronics store sold 48 of the DVD players shown below at the sale price.

How much did the store take in on the sale of the DVD players?

Answers and explanations start on page 202.
For more practice with estimation, see page 166.

Front-End Estimation

When you look at any number, the first digit gives you the most information about the size of the number. For instance, look at the number 5,238. The 5 represents 5 thousand. From this digit alone, you know that this number is at least 5,000 and less than 6,000.

Front-end estimation means using the values of the first digits of numbers to estimate the answer to a problem.

E X A M P L E : Find the sum of 942, 538, and 235.

Add the values of the first digits of each number.

$$
\begin{array}{r}
942 \rightarrow 900 \\
538 \rightarrow 500 \\
+\ 235 \rightarrow +\ 200 \\
\hline
1,600
\end{array}
$$

Compare to the exact answer. The estimate and exact answer are close.

$$
\begin{array}{r}
942 \\
538 \\
+\ 235 \\
\hline
1,715
\end{array}
$$

When using the first digit alone doesn't give a good estimate, use the first two digits.

E X A M P L E : Marcy's high score in a video game was 18,655. Phillips's high score was 11,302. About how many more points did Marcy score than Phillip?

Using the first digit to estimate gives us $10,000 - 10,000 = 0$. This is not a useful estimate. Try the first two digits of both numbers: $18,000 - 11,000 = 7,000$. Marcy scored **about 7,000 more points** than Phillip.

Compatible Numbers

Compatible numbers are numbers that work well together. For instance, you may know that $25 + 75 = 100$. Now suppose you need to add 24 and 73. Since 24 is almost 25 and 73 is close to 75, you could estimate that the sum of 24 and 73 is about 100.

Compatible numbers are very helpful when you need to estimate a quotient.

E X A M P L E : Divide 2,324 by 83.

Use compatible numbers. Think: 2,324 is close to 2,400, and 83 is close to 80.

$$
\begin{array}{r}
30 \\
\text{Estimate: } 80{\overline{\smash{)}2,400}} \\
2,400 \\
\hline
0
\end{array}
$$

$$
\begin{array}{r}
28 \\
\text{Exact answer: } 83{\overline{\smash{)}2,324}} \\
2,324 \\
\hline
0
\end{array}
$$

You have learned three methods of estimation: rounding, front-end estimating, and using compatible numbers. In each situation, use the method that works best for you. Remember, many problems in life and on tests can be solved with a good estimate. Estimating can save you time, and it is one of the best ways to check your work.

A. Show how you would solve the problem by using front-end estimation. Then find the exact answer.

	Estimate	Exact Answer
Example: 854 + 549 + 167	800 + 500 + 100 = 1,400	1,570

1. $24 + $67 + $12 _____ _____
2. 44,914 − 12,350 _____ _____
3. 520 × 23 _____ _____
4. 1,243 ÷ 11 _____ _____

B. Show how you would solve the problem by using compatible numbers. Then find the exact answer.

	Estimate	Exact Answer
Example: 7,440 ÷ 24	7,500 ÷ 25 = 300	310

5. 177 ÷ 59 _____ _____
6. 3,528 ÷ 42 _____ _____
7. 2,226 ÷ 53 _____ _____

C. Estimate an answer to each question using the method that is best for you. Then find the exact answer.

8. Rafi will make 12 payments of $39 to pay for a washing machine. How much will he pay in all for his purchase?

9. A concert hall has three sections. There are 356 seats in the center, 278 seats on the sides, and 186 seats in the balcony. How many seats are there in all?

10. A city budgeted $614,500 for school improvements. The actual cost of the improvements was $869,430. How far over budget was the city?

11. Nine co-workers bought a winning lottery ticket worth $2,025,000. How much is each person's share?

12. Teresa kept track of the ticket sales for the first four showings of a new film.

Ticket Sales by Time	
2:30 show	152 tickets
5:00 show	385 tickets
7:30 show	246 tickets

How many more tickets were sold for the 7:30 show than for the 2:30 show?

Answers and explanations start on page 202.
For more practice with estimation, see page 166.

Decide Which Operation to Use

In the video program, many people use a process for solving complex problems. Once you understand the question and the given information, you can set up how you would solve the problem. This requires thinking about and deciding which operation to use.

Deciding which operation to use means that you need to figure out whether to add, subtract, multiply, or divide to solve a math problem. Here are two different problems that show how to decide which operation is needed.

EXAMPLE 1: Samantha wants to buy four shirts that cost $12 each. How much will the four shirts cost?

Since the cost of all four shirts is the same, use multiplication to solve: $4 \times \$12$.

EXAMPLE 2: There are 15 yards of wire on a spool. If you cut off 9 yards, how many yards are left?

To find the difference between the length of wire you start with and the number of yards cut off, subtract: $15 - 9$.

Problems with Two or More Operations

When using more than one operation to solve a problem, be sure to use the correct order of operations. The **order of operations** tells the order in which to perform calculations, working from left to right.

Order of operations

1. any operations within parentheses
2. powers or roots
3. multiplication and division
4. addition and subtraction

EXAMPLE: Mark drove 300 miles on the first day of his trip and 340 miles on the second day. If he needs to drive a total of 785 miles, how many more miles does he need to travel?

Think about the question: How many more miles does he need to travel?

Find the needed information: How far does he need to go? <u>785 miles</u>
How far has he driven? <u>300 miles and 340 miles</u>

Set up the problem: Decide which operations to use.

Step 1. Add the number of miles driven: $300 + 340$.
Step 2. Subtract the sum from the total number of miles: 785.

Answer: Since you need to subtract the <u>total</u> number of miles driven so far, group the amounts in parentheses so that their sum will be subtracted: $785 - (300 + 340)$.

Set-up Problems

On the GED Math Test, a **set-up problem** will ask you to choose the solution that shows how a problem could be solved. You are not required to find the answer.

> **EXAMPLE:** Jane needs 2 batteries for her remote control, 6 for a flashlight, and 4 for a clock. If Jane buys 2 packages of 8 batteries each, which expression can be used to find how many batteries she will have left?
>
> **(1)** $2 + 6 + 4$
> **(2)** 2×8
> **(3)** $2 \times 8 - (2 + 6 + 4)$
> **(4)** $2 \times 8 + (2 - 6 - 4)$
> **(5)** $8 - (2 + 6 + 4)$

The correct answer is **(3)**. This choice shows to first add the numbers in parentheses (the number of batteries needed). Then multiply to find the total number of batteries in the packages. Last, subtract to find the number of batteries left.

GED MATH PRACTICE

DECIDE WHICH OPERATION TO USE

Choose the <u>one best way</u> to solve each problem. <u>You do not have to find the answer.</u>

1. Eric has 4 bookshelves that hold about 40 books each. About how many books in all can Eric keep on his shelves?
   (1) Multiply 4 by 40.
   (2) Divide 40 by 4.

2. Marta bought 4 cans of soup for $2. How much did each can of soup cost?
   (1) Multiply 4 by $2.
   (2) Divide $2 by 4.

3. Mike earned $62 in tips on Saturday and $55 in tips on Sunday. He earned twice that much last weekend. Which expression can be used to find how much he earned in tips last weekend?
   (1) $62 - 55$
   (2) $2 \times 62 \times 55$
   (3) $2 \times 62 + 55$
   (4) $2 \times (62 + 55)$
   (5) $\frac{62}{55}$

Questions 4 and 5 refer to the drawing.

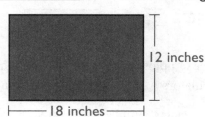

4. How much longer is the picture's length than its width?
   (1) Add: 12 plus 18.
   (2) Subtract: 18 minus 12.

5. Which expression could be used to find the total distance around the edge of the picture?
   (1) $18 + 12$
   (2) $18 - 12$
   (3) 4×12
   (4) $2 \times 12 + 2 \times 18$
   (5) $2 \times 18 - 12$

Answers and explanations start on page 202.
For more practice with estimation, see page 166.

Calculator Operations and Grid Basics

Calculators perform the four basic operations: addition, subtraction, multiplication, and division. The calculator has many more uses, including finding exponents, roots, and solving problems with fractions. These functions will be shown later in this book.

Calculator Operations

On the GED, you will use the Casio *fx*-260 calculator that is shown here. Use the four basic operation keys on this calculator to add, subtract, multiply, and divide. For problems with parentheses, you may want to use the parentheses keys.

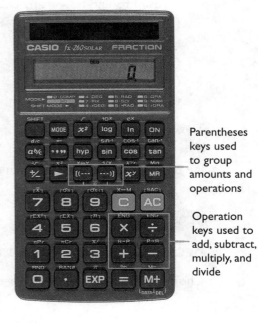

Parentheses keys used to group amounts and operations

Operation keys used to add, subtract, multiply, and divide

Enter each example below on your calculator. Compare your answer to the display shown. Be sure to clear the calculator display before starting a new problem.

EXAMPLES

173 $+$ 246 $=$ [419.]

890 $-$ 369 $=$ [521.]

24 $\times$ 57 $=$ [1368.]

1032 $\div$ 6 $=$ [172.]

The examples below show two or more operations. The Casio *fx*-260 and most other calculators will automatically perform the operations in the correct order.

Enter each example below on your calculator. Compare your answer to the display.

EXAMPLES

8 $+$ 15 $\times$ 7 $=$ [113.]

90 $\div$ 5 $-$ 3 $\times$ 4 $=$ [6.]

These examples show problems with parentheses. For these problems, you must do the operation in parentheses first <u>or</u> use the parentheses keys on your calculator. Try each problem using both methods. Which one is easier for you?

Enter each example below on your calculator. Compare your answer to the display.

EXAMPLES

(11 $+$ 19) $\times$ 5 $=$ [150.]

169 $\div$ (16 $-$ 3) $\times$ 4 $=$ [52.]

Grid Basics

Some problems on the GED Math Test require you to fill in your answer on a grid. You may begin in any column as long as your answer will fit. You might want to get in the habit of always starting your answer in the left column. Write your answer in the boxes on the top row. Fill in the one correct bubble below each number. Leave any unused columns blank.

E X A M P L E :

A concert ticket costs $20. If 4 friends buy tickets together, what is the total cost?

Answer: $20 × 4 = **$80** The answer is shown on the grid.

GED MATH PRACTICE

CALCULATOR OPERATIONS AND GRID BASICS

Use your calculator to solve each problem.

1. 356 + 234 =

2. 9,450 ÷ 50 =

3. 8,200 × 7 =

4. 111 − 93 =

5. 34 + 18 × 5 =

6. (64 − 8) ÷ 7 =

7. Pat jogged 45 minutes each day for 3 days. Then she jogged 30 minutes each day for 2 days. How many minutes in all did Pat jog for the 5 days?

   (1) 60   (4) 225

   (2) 135   (5) 330

   (3) 195

Mark your answer on the grid provided. You may use a calculator.

Questions 8 and 9 refer to the table below.

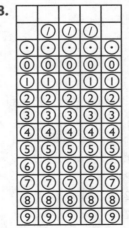

TODAY'S SPECIALS

Strawberries: 2-pound box for $5

Kiwis: 5 for $1

Cherries: $3 per pound

8. Rolando wants to buy 2 pounds of cherries, 5 kiwis, and 3 boxes of strawberries for a party. How much will he spend in all?

9. How much more do 4 pounds of cherries cost than 4 pounds of strawberries?

Answers and explanations start on page 202.
For more practice with calculator operations, see page 167.

GED Review: Problem Solving

Part I: Choose the <u>one best answer</u> to each question below. You <u>may</u> use your calculator.

1. A road crew wants to place a marker every 120 feet along a one-mile stretch of road. If there are 5,280 feet in a mile, how many markers will they need?
    (1) 5,400
    (2) 5,160
    (3) 4,080
    (4) 440
    (5) 44

2. Shau Mei has $250 in her checking account. Her utility bill is $45, and her telephone bill is $37. If she pays her bills and spends another $60 on food, how much will she have left?
    (1) $82
    (2) $108
    (3) $142
    (4) $190
    (5) $392

3. Kevin charges $40 per hour labor plus materials for a service call. If he worked 3 hours on a job and spent $187 for materials, what was the total charge?
    (1) $307
    (2) $227
    (3) $147
    (4) $130
    (5) $120

4. Jason sold 35 adults' tickets for $5 each and 62 children's tickets for $3 each. Which expression could be used to find out how much money he collected?
    (1) $35 + 5 + 62 + 3$
    (2) $35 \times 5 - 62 \times 3$
    (3) $35 \times 5 + 62 \times 3$
    (4) $35 + 62 - (5 \times 3)$
    (5) $35 + 62 \div (5 + 3)$

Questions 5 and 6 refer to the map below.

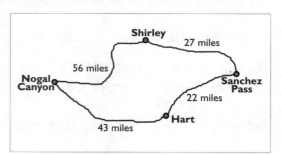

5. How much farther is it from Nogal Canyon to Sanchez Pass through Shirley than through Hart?

6. Janet works for a delivery company in Hart. What is the shortest distance she can travel to make deliveries in Sanchez Pass, Shirley, and Nogal Canyon and return to Hart?

Part II: Choose the <u>one best answer</u> to each question. You <u>may not</u> use your calculator.

Questions 7 and 8 refer to the chart below.

Day	Videos Rented
Monday	57
Tuesday	55
Wednesday	109
Thursday	151
Friday	352
Saturday	566
Sunday	206

7. How many videos were rented for Friday, Saturday, and Sunday?
   - **(1)** 566
   - **(2)** 772
   - **(3)** 918
   - **(4)** 1,124
   - **(5)** 1,496

8. If videos rent for $3 each, how much money did the video store make in all on the two largest rental days?
   - **(1)** $1,698
   - **(2)** $2,316
   - **(3)** $2,754
   - **(4)** $3,372
   - **(5)** $4,488

9. Susan ordered 2 black ink cartridges at $30 each and 1 color cartridge for $35 for her printer. If the shipping cost is $4, which expression could be used to find the total amount of her order?
   - **(1)** 30 + 35 + 4
   - **(2)** 30 × 2 + 35 + 4
   - **(3)** 30 × 2 + 35 × 4
   - **(4)** (30 + 35) × 2
   - **(5)** (30 + 35) ÷ 2

10. Sam's job pays $1,600 per month. He is applying for a job that pays $21,000 a year. How much more will he make EACH MONTH if he gets the new job?
    - **(1)** $150
    - **(2)** $175
    - **(3)** $210
    - **(4)** $1,750
    - **(5)** $1,800

11. If the average shower uses about 35 gallons of water, how many gallons will a family of four use in a week if all family members take a shower every day?
    - **(1)** 140
    - **(2)** 245
    - **(3)** 385
    - **(4)** 700
    - **(5)** 980

12. Rich is putting a new roof on his house. The roof measures 1,200 square feet. One package of shingles covers 100 square feet. If Rich buys two extra packages for breakage, how many packages of shingles will he need?
    - **(1)** 10
    - **(2)** 11
    - **(3)** 12
    - **(4)** 13
    - **(5)** 14

Answers and explanations start on page 203.

Decimals

1. Think About the Topic

The program that you are going to watch is about *Decimals*. The video will explore decimals, the importance of place value, and basic operations with decimals. You will see that we use decimals when we work with money, statistics, and amounts. You will find that the main difference between working with whole numbers and decimals is the use of the decimal point.

2. Prepare to Watch the Video

This program will look at topics ranging from naming and comparing decimals to basic operations with decimals. Answer the questions below to see how your knowledge of money can help you better understand decimals.

A snack costs $2.75. If you pay $3.00, how much change will you get back?

_____

You know you will get a *quarter,* or *$0.25,* in change. Subtracting decimals is like subtracting whole numbers except that you have to line up the decimal points first.

If you owe $10.67 and have all $1 bills but no coins, how much money would you give the cashier?

_____

You would round up because any whole dollar amount that is $10 or less would not cover the cost. If you answered $11, $15, or $20, those are all reasonable amounts to give the cashier.

LESSON GOALS

MATH SKILLS

- Explore decimal place values
- Add and subtract decimals
- Multiply and divide decimals

GED PROBLEM SOLVING

- Does the answer make sense?

GED MATH CONNECTION

- Decimals on the calculator and the grid

GED REVIEW

EXTRA PRACTICE PP. 172–175

- Rounding and Comparing Decimals
- Adding and Subtracting Decimals
- Multiplying and Dividing Decimals
- Calculator Operations with Decimals

3. Preview the Questions

Read the questions under *Think About the Program* below, and keep them in mind as you watch the program. You will be reviewing them after you watch.

4. Study the Vocabulary

Review the terms to the right. Understanding the meaning of math vocabulary will help you understand the video and the rest of this lesson.

WATCH THE PROGRAM

As you watch the program, pay special attention to the host who introduces or summarizes major ideas that you need to learn about. The host may also tell you important information about the GED Math Test.

decimal—a number that represents part of a whole; written using digits after a decimal point; for example, 0.25

dividend—the number you divide in a division problem

divisor—the number you divide by in a division problem

placeholder—a zero used to fill a place-value column

quotient—the answer to a division problem

regroup—to carry an amount to another place value

AFTER YOU WATCH  22

1. Think About the Program

What are three ways decimals are used in daily life?

How do you name a decimal? For example, how would you say .89?

How do you compare decimals? For example, how would you compare .04, .4, and .004 to see which is the largest?

Which operation (adding, subtracting, multiplying, or dividing) requires you to count the number of decimal places in the problem before putting the decimal point in the answer?

2. Make the Connection

The program talked about keeping track of money in a checking account. What are the advantages of having a checking account? Do you have one? How do you keep track of your account—with paper and pencil or a calculator?

"The power of decimals is that they enable fractions to look like whole numbers."

Decimal Place Values

The Meaning of Decimal Values

Decimals show values that are less than one whole. When you see a price tag for $2.25, you know that the price means 2 whole dollars and 25 cents. The decimal part of the price, $.25, is a fractional part of a whole dollar.

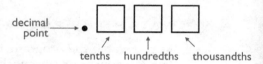

Look at the chart to the right. In a decimal, the placement of a digit tells you its value. To better understand this idea, study the number lines that follow.

EXAMPLE: Point *A* is located between 2 and 3 on the number line. What is the value of point *A*?

Now we move in closer. This number line divides the space between 2 and 3 into ten equal spaces called *tenths*. Point *A* is between 2.6 and 2.7 (*2 and 6 tenths* and *2 and 7 tenths*).

Moving in still closer, we can divide the space between 2.6 and 2.7 into ten equal spaces called *hundredths*.

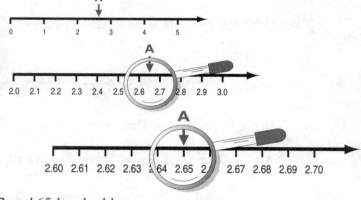

Answer: Point *A* is located at **2.65,** or *2 and 65 hundredths*.

Reading and Writing Decimals

Follow these steps to express a decimal in words.

Step 1. Read the whole number part.
Step 2. Say "and" for the decimal point.
Step 3. Read the digits to the right of the decimal point as you would a whole number.
Step 4. Say the name of the place-value column of the last decimal digit.

EXAMPLES   0.6 is read *six tenths*.
1.24 is read *one and twenty-four hundredths*.
0.432 is read *four hundred thirty-two thousandths*.
15.008 is read *fifteen and eight thousandths*.

An informal way to read a decimal is to say "point" for the decimal point. The number 3.15 becomes *"three point one five."*

You will sometimes need to use zeros as placeholders when you write numbers.

EXAMPLE: Write *five and sixteen thousandths* using digits.

The whole number part of the number is 5. The word *and* represents the decimal point. The decimal part of the number (16) must end in the thousandths column. Since the number 16 has only two digits, put a placeholder zero in the tenths column to "push" the 16 out to the thousandths column.

5.016
↑
placeholder
zero

Answer: Five and sixteen thousandths is written **5.016.**

DECIMALS ▪ PRACTICE 1

A. Answer the following questions.

1. In the number 18.394, which digit is in the
   a. hundredths place? _____
   b. ones place? _____
   c. tenths place? _____

2. In the number 206.175, which digit is in the
   a. hundredths place? _____
   b. ones place? _____
   c. tenths place? _____

B. Write each number in words.

Example: 3.12   *three and twelve hundredths*

3. 4.9 _____

4. 0.125 _____

5. 16.018 _____

6. 0.32 _____

7. 7.509 _____

8. 50.008 _____

C. Write each number using digits.

Example: Thirty and five hundredths _30.05_

9. eight and six tenths _____

10. seventy-five thousandths _____

11. ten and fifty-seven hundredths _____

12. one and thirty-six hundredths _____

13. six and eighty-one thousandths _____

14. eleven and nine tenths _____

Answers and explanations start on page 203.

Comparing and Ordering Decimals

Would you rather earn $12.50 an hour or $8.75? Obviously, you would choose $12.50 because $12 is more than $8. When numbers have different whole number parts, you don't need to think about the decimals to decide which is larger.

Sometimes you will need to look at the decimal part of the numbers to compare sizes. To help you put the numbers in the same form, you can add zeros at the end of a decimal without changing its value.

EXAMPLE: The diameters of three wires are 0.35, 0.4, and 0.125 centimeter. Arrange the diameters in order from least to greatest.

Step 1. Write the numbers in a column, and align the decimal points.   0.350
Step 2. Write zeros to the right of the last decimal digit so that each   0.400
number has the same number of decimal digits. Adding zeros   0.125
at the <u>end</u> of a decimal number <u>will not change</u> its value.
Step 3. Compare the decimals as you would whole numbers: 350, 400, and 125.

Answer: From least to greatest, the measurements are **0.125, 0.35,** and **0.4 centimeters.**

Rounding Decimals

A store employee uses a calculator to find the sales tax on a purchase. After multiplying, she sees the calculator display at right. How would you change the display to dollars and cents?

You need an amount with only two decimal places. Round to the nearest cent. The cents column is the hundredths place: 1.45<u>8</u>5. Since the digit to the right of the hundredths place is greater than 5 (more than half), the employee should round up to $1.46.

Rounding decimals is similar to rounding whole numbers.
- Find the place value to which you need to round.
- Look at the digit to the right of that place value. If the digit is 5 or more, round up. If the digit is less than 5, round down.
- Drop any decimal places to the right of the place value that you rounded to.

EXAMPLES

Round 3.562 to the nearest <u>tenth</u>.   3.<u>5</u>72 rounds up to **3.6**
Round 0.23 to the nearest <u>tenth</u>.   0.<u>2</u>3 rounds down to **0.2**
Round 5.946 to the nearest <u>hundredth</u>.   5.9<u>4</u>6 rounds up to **5.95**
Round $1.024 to the nearest <u>cent</u>. $1.0<u>2</u>4 rounds down to **$1.02**
Round 0.87 to the nearest <u>whole number</u>.   <u>0</u>.87 rounds up to **1**

A. Round each number as directed.

1. Round 15.684 to the nearest tenth.

2. Round $4.126 to the nearest cent.

3. Round 3.74 to the nearest whole number.

4. Round 0.4917 to the nearest hundredth.

5. Round 16.147 to the nearest tenth.

6. Round $10.37 to the nearest dollar.

7. Round 0.506 to the nearest hundredth.

8. Round $0.975 to the nearest cent.

B. Compare each pair of numbers. Write <, >, or = in the blank. Refer to page 24 for a review of the comparison symbols.

9. 14.9 _____ 19.4

10. 1.65 _____ 1.63

11. 2.80 _____ 2.8

12. 0.425 _____ 0.325

13. 3.56 _____ 3.561

14. 0.783 _____ 0.739

15. 0.045 _____ 0.05

16. 8.315 _____ 7.915

17. 4.60 _____ 4.6

18. 10.032 _____ 10.302

C. Use rounding, comparing, and ordering to answer the following questions.

19. Arrange the following in order from greatest to least: $46.50, $42.60, $46.60.

20. A chemist weighs a compound on a digital scale. The scale reads 5.826 grams. What is the weight to the nearest tenth gram?

21. The times for 4 runners are shown below.

James	5.16 minutes
Roy	4.525 minutes
Armando	4.168 minutes
Ken	4.52 minutes

Write the names of the runners in the order they finished the race. (*Hint:* The runner who finished first ran the race in the least amount of time.)

22. Library books are organized using a decimal system. Each book has a call number. A sign states that the range of call numbers on a shelf is 731.18–805.95. Which of the books below belong on this shelf?

Book A — 750.04
Book B — 729.5
Book C — 802.96
Book D — 731.18

23. In Canada, the end zone of a football field is 59.47 by 22.87 meters. Write both measurements rounded to the nearest whole meter.

Answers and explanations start on page 203.
For more practice rounding and comparing decimals, see page 168.

MATH SKILLS

"Decimals are the easiest form of fractions to add and subtract."

MATH SKILLS

Adding and Subtracting Decimals

Addition

When you add decimals, you regroup just as you do when adding whole numbers. **Regrouping** works because decimals and whole numbers both use place value.

EXAMPLE: On a trip, Vanessa bought 11.6 gallons of gasoline. At her next two stops, she purchased 12.5 gallons and 13 gallons. How many gallons of gas did she buy?

Step 1. Write the numbers in a column, lining up the decimal points.

Step 2. If needed, write zeros to the right of the last decimal digit in a number so that the numbers you are adding have the same number of decimal places.

Step 3. Start on the right and add as you would with whole numbers. Regroup to the next place-value column as needed.

Step 4. Place the decimal point in the answer directly below the decimal points in the problem.

$$\begin{array}{r} 11.6 \\ 12.5 \\ + 13.0 \\ \hline 37.1 \end{array}$$

Answer: Vanessa bought **37.1 gallons** of gas on her trip.

Estimating

Estimating before you add will help you check that you correctly placed the decimal point in your answer. To estimate decimals, you can round them.

EXAMPLE: Add: 0.7 + 1.325 + 3.15

To estimate, round each amount to the nearest whole number. Add to estimate an answer: 1 + 1 + 3 = 5.

The answer should be close to 5.

0.7 rounds up to 1
1.325 rounds down to 1
3.19 rounds down to 3

Step 1. Write the numbers in a column. Line up the decimal points.

Step 2. Write zeros so that each number has the same number of decimal places. Add and bring down the decimal point.

$$\begin{array}{r} 0.700 \\ 1.325 \\ + 3.150 \\ \hline 5.175 \end{array}$$

Compare the total to the estimate. **5.175** is close to the estimate of 5.

A. Solve.

1. $\begin{array}{r} 2.5 \\ +\ 6.8 \\ \hline \end{array}$

2. $\begin{array}{r} 4.9 \\ +\ 0.6 \\ \hline \end{array}$

3. $\begin{array}{r} 2.58 \\ +\ 5.61 \\ \hline \end{array}$

4. $\begin{array}{r} \$27.57 \\ +\ \ \ 8.94 \\ \hline \end{array}$

5. $\begin{array}{r} 6.7 \\ 14.6 \\ +\ 0.9 \\ \hline \end{array}$

6. $\begin{array}{r} 4.106 \\ 0.425 \\ +\ 2.609 \\ \hline \end{array}$

B. Rewrite each problem in a column and solve.

7. 3.42 + 4.6 + 0.15 =

8. $1.36 + $0.98 =

9. 0.907 + 9.7 + 0.79 =

10. 12.5 + 0.136 + 0.78 =

11. 0.9 + 1.86 + 2.4 =

12. 134.6 + 27 + 5.45 =

C. Solve.

13. The total rainfall for March was 6.48 inches. The rainfall for April and May was 5.6 and 4.125 inches. What was the total rainfall for the three-month period?

14. To make a tabletop, Melanie glues three layers of wood together. The layers are shown below. What is the total thickness in centimeters of the tabletop?

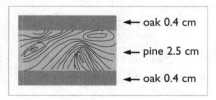

← oak 0.4 cm
← pine 2.5 cm
← oak 0.4 cm

15. Janice bought a sandwich for $6.95 and a drink for $1.27. What was the total cost of the items?

16. Chris ran 2.7 miles on Monday, 3.6 miles on Wednesday, and 1.8 miles on Friday. How many miles did he run in all?

17. Matt plans to put in a wood fence on three sides of his backyard. The longest side is 24.5 meters long. The remaining sides are 13.25 and 15.625 meters. What is the total length in meters that Matt needs to fence?

18. A catalog lists the following shipping weights for three items:

twin bed comforter	5.44 pounds
twin sheets	2.5 pounds
matching curtains	1.85 pounds

What is the total weight in pounds of the three items?

Answers and explanations start on page 203.
For more practice adding decimals, see page 169.

MATH SKILLS

Subtraction

Subtract decimal numbers the same way that you subtract whole numbers.

EXAMPLE: A metal rod is 3.8 meters long. If a piece measuring 1.675 meters is cut from the rod, how many meters are left?

3.8 meters

Step 1. Write the numbers in a column, lining up the decimal points.

Step 2. If needed, write zeros to the right of the last decimal digits so that the numbers have the same number of decimal places.

Step 3. Start on the right and subtract as you would with whole numbers. Regroup as needed.

Step 4. Place the decimal point in the answer directly below the decimal points in the problem.

$$\begin{array}{r} 3.800 \\ -1.675 \\ \hline \end{array} \qquad \begin{array}{r} {\scriptstyle 7\,9\,10} \\ 3.8\cancel{0}\cancel{0} \\ -1.675 \\ \hline \mathbf{2.125} \end{array}$$

Answer: There are **2.125 meters** left over.

Estimating

Estimating is a good way to make sure your answer is reasonable. In the problem above, you could think 3.8 rounds to 4 and 1.675 rounds to 2. Subtract: $4 - 2 = 2$. Compare your answer to the estimate. Since 2.125 is close to 2, the answer is reasonable.

If you need to be certain your answer is correct, use addition to check subtraction. Add the answer to the number you subtracted. The sum should be the number you started with. *Note:* $3.800 = 3.8$.

$$\begin{array}{r} {\scriptstyle 1\ 1} \\ 2.125 \\ -1.675 \\ \hline \mathbf{3.800} \end{array}$$

Two-Step Problems

Some problems involve both addition and subtraction. When problems have only addition and subtraction, do the operations in the order that seems easiest to you.

EXAMPLE: Linda has a bank balance of $189.45. She deposits $367.90 and writes a check for $136.18. After these transactions, what is her new bank balance?

Deposits are added to bank accounts, and checks are subtracted. Does it matter which you do first? Not at all. As long as you have an amount large enough to cover the amount being subtracted, the order won't affect the answer.

Method 1: Add the deposit first.

$$\text{Add: } \begin{array}{r} \$189.45 \\ +\,367.90 \\ \hline \$557.35 \end{array} \qquad \text{Subtract: } \begin{array}{r} \$557.35 \\ -\,136.18 \\ \hline \$421.17 \end{array}$$

Method 2: Subtract the check first.

$$\text{Subtract: } \begin{array}{r} \$189.45 \\ -\,136.18 \\ \hline \$\,53.27 \end{array} \qquad \text{Add: } \begin{array}{r} \$\,53.27 \\ +\,367.90 \\ \hline \$421.17 \end{array}$$

MATH SKILLS

A. Solve.

1.
```
   0.94
 − 0.48
```

2.
```
   4.72
 − 1.92
```

3.
```
  15.126
 − 4.705
```

4.
```
   25.7
 − 6.85
```

5.
```
   7.53
 − 6.9
```

6.
```
  18.04
 − 3.125
```

7.
```
   50
 − 13.75
```

8.
```
   12.5
 − 0.48
```

9.
```
  36.008
 − 15.9
```

B. Rewrite each problem in a column and solve. Check using addition.

10. 11.5 − 3.68 =

11. $18 − $5.34 =

12. 75.005 − 4.25 =

13. $103.87 − $34.50 =

14. 1.5 − 0.009 =

15. 18.369 − 12.4 =

C. Solve.

16. Candice read her car's odometer at the beginning and end of the summer. The odometer readings are shown below.

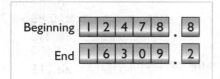

How many miles did she drive her car during the summer?

17. One bolt is 1.275 inches long. A second bolt is 0.8 inches long. How much longer is the first bolt than the second bolt?

18. Max paid for a $27.92 purchase with two $20 bills. How much change did he receive?

19. Jesse has $1,346.87 in his checking account. He writes checks for $538.70 and $45 and deposits a check for $126.42. How much money does he have in his account after the transactions?

20. Normal body temperature is 98.6 degrees. Stuart has a temperature of 103.2 degrees. How many degrees above normal is Stuart's temperature?

21. CFH stock gained 1.4 points per share on Monday, 4.8 on Tuesday, and 2.5 on Wednesday. During the same three-day period, KRO stock increased 9 points per share. How many more points did KRO stock gain than CFH?

Answers and explanations start on page 204.
For more practice subtracting decimals, see page 169.

"Place value is the key to the systems on both sides of the decimal point."

Multiplying and Dividing Decimals

Multiplication

When you multiply, the number of decimal places in the answer is equal to the number of decimal places in the problem.

EXAMPLE: Find the product of 2.5 and 0.7.

$$\begin{array}{r} {}^{3} \\ 2.5 \\ \times\ 0.7 \\ \hline \mathbf{1.75} \end{array}$$

Step 1. Multiply as you would with whole numbers.

Step 2. Place the decimal point in the answer. There are two decimal places in the problem. Count two places from the right, and place the decimal point in the product.

The answer 1.75 is smaller than 2.5, the number you started with. Are you surprised? When you multiply by 1, the number doesn't change: $4 \times 1 = 4$. When you multiply by a number greater than 1, the answer is greater than the number you multiplied: $4 \times 2 = 8$. Therefore, when you multiply by a value less than 1, the answer will be less than the number you multiplied.

When counting the number of decimal places in the answer, you may have to add zeros to the left of the product to place the decimal point.

EXAMPLE: Multiply: 0.105×0.83.

$$\begin{array}{r} 0.105 \leftarrow 3 \text{ places} \\ \times\ 0.83 \leftarrow 2 \text{ places} \\ \hline 315 \\ 8400 \\ \hline \mathbf{.08715} \leftarrow 5 \text{ places} \end{array}$$

Step 1. Multiply as you would whole numbers.

Step 2. Count the number of decimal places in the problem. There are 3 in the top number and 2 in the bottom number, for a total of 5. Count 5 places from the right in the answer.

Step 3. Since there are only 4 digits in the answer, add a placeholder zero to the left before placing the decimal point.

Rounding Decimal Answers

If a product has more decimal places than you need, you must round to the appropriate place-value column. For example, when you work with money, your answer should be rounded to the nearest cent.

EXAMPLE: Find the sales tax on a purchase of $20.50 by multiplying by 0.085.

Step 1. Multiply as you would whole numbers.

Step 2. Place the decimal point in the answer.

Step 3. Round to the nearest cent. 1.74250 rounds to $1.74.

$$\begin{array}{r} \$20.50 \\ \times\ 0.085 \\ \hline 10250 \\ 164000 \\ \hline \mathbf{\$1.74250} \end{array}$$

Answer: The sales tax is **$1.74**.

A. Solve.

1. 3.5
    × 4

2. 7.3
    × .2

3. 0.08
    × .5

4. $5.60
    × 1.5

5. 84
    × 4.2

6. $10.12
    × 0.6

7. 2.012
    × 0.25

8. 0.082
    × 0.07

9. $28.69
    × 0.07

B. Rewrite each problem in a column and solve.

10. 4.82 × 0.6 =

11. 0.0069 × 5 =

12. 0.354 × 9 =

13. 124.6 × 0.005 =

14. 4.006 × 25 =

15. 2.8 × 0.43 =

C. Solve.

16. Cyndi makes monthly payments of $180.65 on her new car. How much will she pay in one year? (*Hint:* 1 year = 12 months)

17. A container of water weighs 7.6 pounds. How many pounds would 9 containers weigh?

18. John's work pays him $0.35 for each mile he drives his car for work. If he drives 82.5 miles in one week, how much will he be paid for mileage?

19. Rafael has a credit line with the bank. To find the amount of interest he must pay each month, the bank multiplies the amount he owes by 0.0175. How much interest will Rafael pay if he owes $900?

Questions 20 through 22 refer to the table below.

West Trails Supplies	
Catalog Item	**Shipping Weight**
Hiking boots	2.64 pounds
Backpacking frame	4.2 pounds
3-person dome tent	8.125 pounds

20. A scout group buys 4 tents. What is the shipping weight of the order?

21. Eight boys order backpacking frames from the company. What is the total shipping weight of their order?

22. Three leaders each buy a backpacking frame and a pair of hiking boots. What is the total shipping weight of their orders?

Answers and explanations start on page 204.
For more practice multiplying decimals, see page 170.

Dividing Decimals

To divide decimals, use the same process that you use for whole numbers. The only difference is that you must figure out where the decimal point goes in the answer before you divide.

Learning a few terms will make it easier to understand this process. In a division problem, the number you are dividing is called the **dividend.** The **divisor** is the number you are dividing by, and the **quotient** is the answer to the problem.

Study the three different situations below, and make sure you understand where to place the decimal point in the quotient.

EXAMPLE: 27.2 ÷ 4

- The divisor has no decimal places, but there are decimal places in the dividend.
- Place the decimal point in the quotient directly above the decimal point in the dividend.

```
      6.8
  4) 27.2
     24
      3 2
      3 2
```

EXAMPLE: 14.05 ÷ 0.5

- There are decimal places in the divisor and the dividend.
- Move the decimal point in the divisor to the right to make it a whole number.
- Move the decimal point in the dividend the same number of places.
- Place the decimal point in the quotient directly above the decimal point in the dividend.

```
       28.1
  .5) 14.0.5
      10
      40
      40
       5
       5
```

EXAMPLE: 7.2 ÷ 0.04

- There are more decimal places in the divisor than there are in the dividend.
- Move the decimal point in the divisor to the right to make it a whole number.
- Add zeros to the dividend so that you can move the decimal point the same number of places.
- Place the decimal point in the quotient.

```
       1 80.
  .04) 7.20.
       4
       3 2
       3 2
```

Some division problems continue longer than you expected. If a quotient has more decimal digits than you need, round your answer. If the problem is about money, round to the nearest cent.

```
        1.687
  8) 1 3.500
     8
      5 5
      4 8
       70
       64
        60
        56
         4
```

EXAMPLE: Divide $13.50 by 8.

To round to the nearest cent, continue dividing until you have three decimal places. Then stop dividing and round. The quotient 1.687 rounds to **$1.69.**

A. Solve.

1. $7\overline{)25.2}$

4. $0.8\overline{)1.728}$

7. $0.8\overline{)484}$

2. $9\overline{)11.25}$

5. $3.6\overline{)0.324}$

8. $1.2\overline{)576}$

3. $6\overline{)19.08}$

6. $0.25\overline{)55.75}$

9. $1.05\overline{)27.3}$

B. Solve. Round to the nearest cent.

10. Divide $21 by 9.

12. Divide $1,000 by 13.

11. Divide $5.50 by 7.

13. Divide $2.33 by 4.

C. Solve.

14. Andy worked nine hours and earned $113.04. How much was he paid per hour?

15. A barrel holds 24 liters of sports drink. How many 1.5-liter bottles can be filled from the barrel?

16. In six days, Carolina drove 74.7 miles. If she drove about the same number of miles each day, how many miles did she drive a day? Round your answer to the nearest tenth mile.

17. Nita bought a roast that weighs 5.2 pounds. If she paid $41.34, what was the price of the roast per pound?

Questions 18 and 19 refer to the following information.

Garden Center Sale	
3-inch plants	$1.29 each
5-inch plants	$3.79 each

18. Monica has $20 to spend on plants. How many 3-inch plants can she buy?

19. The Garden Center paid $500 for a load of 6-inch plants. If 200 plants were shipped, how much did each plant cost?

Answers and explanations start on page 204.
For more practice dividing decimals, see page 170.

Does the Answer Make Sense?

In the video, you saw many people using decimals in everyday life. The main challenge in working with decimals is correctly placing the decimal point. You can avoid common mistakes with decimals by checking to see that your answer makes sense.

Does the answer make sense? Remember, checking your answer means checking your math _and_ checking to see that your answer is sensible.

On the GED Math Test, you will need to think about your answers to catch any errors in your choice of operations or to spot a misplaced decimal point.

EXAMPLE 1: Rhonda buys a sweater for $19.85. If the sales tax is $1.69, what is the total amount of her purchase?

(1) $18.16
(2) $21.54

Which answer makes sense?

(2) $21.54 The amount of the purchase is close to $20, and the amount of the tax is less than $2. Your number sense tells you that the answer is between $20 and $22.

Option (1) results from incorrectly subtracting the two numbers.

EXAMPLE 2: The length of Lyle's living room is 17.5 feet, and the width is 20.5 feet. How many square feet of carpeting will he need to cover the floor?

(1) 358.75
(2) 3,587.5

Which answer makes sense?

(1) 358.75 Both the width and the length are about 20 feet. The answer should be about 20 × 20, which is 400 square feet.

Option (2) results from incorrect placement of the decimal point.

Estimating with Decimals

Rounding decimals to the nearest whole number is an easy way to estimate the answer before you work the problem.

EXAMPLE: Kevin worked 42 and one-quarter hours on Monday, Tuesday, and Wednesday. How many hours did he work in all?

Estimate the answer using whole hours. (_Hint:_ 42 and one-quarter hours = 42.25 hours) Since 42.25 rounds to 40, estimate 40 × 3 = 120 hours.

Exact Answer: The answer should be a little more than 120 hours.

$$
\begin{array}{r}
42.25 \\
42.25 \\
+\ 42.25 \\
\hline
126.75
\end{array}
$$

The answer 126.75 makes sense.

Multiple-Choice Questions

Sometimes you can make multiple-choice problems easier by eliminating answer choices that don't make sense.

E X A M P L E : Marie orders the dinner special for $6.75. If the tax is $0.55, how much change will she receive from $20?

(1) $7.30   In this problem, options (4) and (5) can be eliminated because

(2) $12.70   the change Marie will receive must be less than $20.

(3) $13.25

(4) $26.75

(5) $27.30   The correct answer is **(2) $12.70.**

GED MATH PRACTICE

DOES THE ANSWER MAKE SENSE?

Estimate to see which answer makes sense. Then solve the problem.

1. If BBQ ribs are $4.95 a pound, how much will Brian pay for 3.5 pounds?
   (1) $17.33
   (2) $8.45

2. If the total bill for dinner is $38.40, how much will Jason and his 3 friends pay if the bill is evenly split?
   (1) $12.80
   (2) $9.60

3. Marika's car gets 18.5 miles per gallon in city driving and 26.8 miles per gallon in highway driving. What is the difference in miles per gallon?
   (1) 8.3
   (2) 45.3

4. The contents of a jar of baby food weigh 5.8 ounces. How many ounces are in a dozen jars?
   (1) 58
   (2) 69.6

Solve each problem. Eliminate as many options as possible before solving.

5. One route to Kent is 35.9 miles. A second route is 47.2 miles. How much longer is the second route than the first?
   (1) 83.1
   (2) 47.2
   (3) 35.9
   (4) 15.0
   (5) 11.3

6. A figure has sides measuring 9.9 cm, 11.2 cm, 8.7 cm, and 12 cm. What is the distance around the figure?
   (1) 21.1
   (2) 29.8
   (3) 31.9
   (4) 41.8
   (5) 118.8

Answers and explanations start on page 204.

Decimals on the Calculator and Grid

The calculator can be used to add, subtract, multiply, and divide decimals. When you work with decimals, be careful to enter the decimal point in the correct place. If you do, the calculator will correctly place the decimal point in the answer.

Calculator Operations

To enter decimals on the Casio *fx*-260, press the decimal point key, $\boxed{\cdot}$. Enter it in the same position as shown in the number. With money, enter the whole number, the decimal point, and the cents. Ignore the dollar sign.

For example, to enter $197.35:

Enter the digits in the whole number part from left to right.
The display shows: $\boxed{\qquad 197.}$

Next press the decimal point key: $\boxed{\cdot}$
The display does not change until you enter the next digit.

Enter the decimal portion of the number from left to right.
The display shows: $\boxed{\qquad 197.35}$

$\boxed{\cdot}$ Decimal Key

When you have a decimal number with one or more zeros on the end of the decimal portion, you do not have to enter the **trailing zeros.** You must, however, enter all other zeros. These **placeholder zeros** affect the place value of digits.

EXAMPLES: Enter each number on your calculator, and compare your display with the one shown. Be sure to clear your calculator before starting the next example.

Trailing zeros may be dropped: $3.40 $\boxed{\qquad 3.4}$ .910 $\boxed{\qquad 0.91}$

Placeholder zeros must be entered: $40.89 $\boxed{\qquad 40.89}$ 1,204.75 $\boxed{\qquad 1204.75}$

The examples below show operations with decimals.

EXAMPLES: Enter each problem on your calculator. Compare your answer with the display

$500.00 $\boxed{+}$ $35.00 $\boxed{=}$ $\boxed{\qquad 535.}$ 60.5 $\boxed{\times}$ 2 $\boxed{=}$ $\boxed{\qquad 121.}$

$20 $\boxed{-}$ $13.79 $\boxed{=}$ $\boxed{\qquad 6.21}$ 30 $\boxed{\div}$ 4 $\boxed{=}$ $\boxed{\qquad 7.5}$

5.04 $\boxed{+}$ 7.26 $\boxed{\times}$ 3 $\boxed{=}$ $\boxed{\qquad 26.82}$

Grid Basics

To fill in a decimal answer on a grid, write each digit and the decimal point in the top row of boxes. Each digit and the decimal point have their own box. Then fill in the matching bubbles below.

EXAMPLE: Joe spends $2.75 on a shake. How much change will he get from $10.00?

Subtract: $10\ \boxed{-}\ 2.75\ \boxed{=}\ \boxed{\qquad 7.25}$ The answer is marked on the grid. Note that there is no dollar sign on the grid.

7	.	2	5	

GED MATH PRACTICE

DECIMALS ON THE CALCULATOR AND GRID

Use your calculator to solve each problem below.

1. $5.6 + 2.04 =$

2. $\$400.00 \div 3 =$

3. $9.5 \times 8.1 =$

4. $12 - 3.7 =$

5. $4.54 + 2.6 \times 5 =$

6. $(4 - 0.8) \div 4 =$

Mark your answer on the grid provided.

Questions 7 and 8 refer to the table below.

Casey's Cabins Nightly Rental Rates and Taxes			
	Rate	State Tax	Hotel Tax
Cabin A	$79.00	$5.53	$3.16
Cabin B	$89.00	$6.23	$3.56
Cabin C	$99.00	$6.93	$3.96

7. What is the total cost including taxes for renting Cabin B for one night?

8. How much more are the combined taxes for Cabin C than Cabin A?

7.

8.

Answers and explanations start on page 205.

GED Review: Decimals

Part I: Choose the <u>one best answer</u> to each question below. You <u>may</u> use your calculator.

1. A case of 24 16-ounce bottles of drinking water is on sale for $11.95. What is the cost for 1 bottle, rounded to the nearest cent?
   (1) $0.45
   (2) $0.50
   (3) $0.55
   (4) $0.67
   (5) $0.75

<u>Questions 2 and 3</u> refer to the table below.

Simpson's Movie Theater: All tickets $6.50			
	Small	Medium	Large
Popcorn	$2.25	$3.00	$3.75
Drinks	$1.50	$2.00	$2.50

2. Roberta and Angela bought 2 movie tickets, a large popcorn, and 2 medium drinks. How much did they spend?
   (1) $14.25
   (2) $16.75
   (3) $20.75
   (4) $27.50
   (5) $41.50

3. For groups of 12 or more, Simpson's Movie Theater charges only $10 per person for a ticket, medium popcorn, and medium drink. How much is saved per person over the regular price?
   (1) $11.50
   (2) $6.50
   (3) $5.00
   (4) $1.50
   (5) $0.75

4. Rebecca makes $7.25 an hour working at a gift shop. If she works 6.5 hours a day for 5 days, how much money will she earn?
   (1) $235.63
   (2) $217.50
   (3) $145.00
   (4) $83.38
   (5) $47.13

5. Joe lost an average of 2.5 pounds each week for 13 weeks. How many pounds did Joe lose in all?
   (1) 2.5
   (2) 5.2
   (3) 13.0
   (4) 15.5
   (5) 32.5

6. A map scale shows 1 inch = 30 miles. How many miles do 5.25 inches represent?
   (1) 1,575
   (2) 157.5
   (3) 15.7
   (4) 1.57
   (5) 0.157

7. At one supermarket, potatoes cost $0.39 per pound. If a 10-pound bag costs $2.79, how much money will you save if you buy the 10-pound bag? Round to the nearest cent.

Part II: Choose the <u>one best answer</u> to each question below. You <u>may not</u> use your calculator.

Questions 8 and 9 refer to the chart below.

Gloria's Running Time for 1 Mile This Week	
	Time in Minutes
Monday	5.71
Tuesday	5.53
Wednesday	4.99
Thursday	5.15
Friday	5.20

8. List Gloria's running times from fastest to slowest (least to greatest).
  (1) 5.71, 5.53, 4.99, 5.15, 5.20
  (2) 5.71, 5.53, 5.20, 5.15, 4.99
  (3) 4.99, 5.15, 5.20, 5.71, 5.53
  (4) 4.99, 5.15, 5.20, 5.53, 5.71
  (5) 5.15, 5.20, 5.53, 5.71, 4.99

9. Gloria's best time for last week was 5.04 minutes. How many minutes faster was her best time this week?
  (1) 5.0
  (2) 0.5
  (3) 0.05
  (4) 0.005
  (5) 0.0005

10. Markus drove 3.5 hours at 60 miles per hour and 1.5 hours at 30 miles per hour through a construction zone. Which expression could be used to find the total number of miles traveled?
  (1) 60 + 3.5 + 30 + 1.5
  (2) (60 + 30) × (3.5 + 1.5)
  (3) 60 × (3.5 + 1.5) × 30
  (4) 3.5 × (60 + 30) × 1.5
  (5) 3.5 × 60 + 1.5 × 30

11. A package of 500 sheets of paper for Tami's printer costs $7.50. What is the cost per sheet?
  (1) $1.50
  (2) $0.15
  (3) $0.015
  (4) $0.0015
  (5) $0.00105

12. For every dollar that Jim saves for college, his parents will contribute $1.50. If Jim saves $500 this year, how much will he have in all after his parents' contribution?
  (1) $12,500
  (2) $7,500
  (3) $1,250
  (4) $750
  (5) $500

13. Rosa counted 315 quarters in her change jar. How much is this in dollars and cents?
  (1) $1,260.00
  (2) $340.00
  (3) $290.00
  (4) $78.75
  (5) $12.60

14. Steve's car gets about 25 miles a gallon. If his car's gas tank holds 17.5 gallons, about how many miles can Steve drive on a full tank of gas?

Answers and explanations start on page 205.

Fractions

LESSON GOALS

MATH SKILLS

- Simplify and compare fractions
- Add and subtract fractions
- Multiply and divide fractions

GED PROBLEM SOLVING

- Find the needed information

GED MATH CONNECTION

- Fractions on the calculator and grid

GED REVIEW

EXTRA PRACTICE pp. 172–175

- Compare Fractions
- Add and Subtract Fractions
- Multiply and Divide Fractions
- Calculators and Grids with Fractions

1. Think About the Topic

The program that you are going to watch is about *Fractions*. The video will show you the meaning of different types of fractions and how to work with them. You will see people using fractions when measuring amounts. Developing a sense for the size of different fractions will help you solve many different types of problems.

2. Prepare to Watch the Video

This program will give ideas about working with fractions. Answer the questions below to try out your fraction sense.

What is half of 200? _____

You should have written 100. When you find $\frac{1}{2}$ of an amount, you're really multiplying by $\frac{1}{2}$. Notice that multiplying a whole number by a fraction ended in an amount that is less than the original amount.

One recipe calls for $\frac{1}{3}$ cup of butter, and a second recipe calls for $\frac{1}{4}$ cup of butter. Which recipe calls for more butter—the first recipe or the second?

_____

The first recipe calls for more butter since $\frac{1}{3}$ of something is larger than $\frac{1}{4}$. That might seem confusing because 4 is larger than 3, but $\frac{1}{3}$ of a stick of butter is larger than $\frac{1}{4}$.

"Picturing" the size of fractions helps you to work with them on the GED and in your everyday life.

3. Preview the Questions

Read the questions under *Think About the Program* below and keep them in mind as you watch the program. You will be reviewing them after you watch.

4. Study the Vocabulary

Review the terms to the right. Understanding the meaning of math vocabulary will help you understand the video and the rest of this lesson.

WATCH THE PROGRAM

As you watch the program, pay special attention to the host who introduces or summarizes major ideas that you need to learn about. The host may also tell you important information about the GED Math Test.

AFTER YOU WATCH

1. Think About the Program

What are some different ways of using fractions in daily life?

What types of measuring tools rely on fractions?

A fraction is another way to write which operation? What does the fraction bar mean?

2. Make the Connection

The program talked about math as a life skill. For example, 1 quarter is $\frac{25}{100}$ of a whole. Therefore, a quarter is \$0.25, or $\frac{1}{4}$ of a dollar. Did you notice that it takes 4 quarters to make a whole? What other things in daily life are divided into fourths? What things in daily life are divided into fourths or halves? How many halves does it take to make a whole? Do you notice a pattern?

common denominator— a number that each denominator in the problem can divide into evenly

denominator—the bottom number in a fraction; it shows the total number of parts in the whole

equivalent fractions— different fractions that have the same value; for example, $\frac{1}{2}$ and $\frac{3}{6}$

improper fraction—a fraction with a value equal to or greater than 1; its numerator is equal to or greater than its denominator, as in $\frac{5}{5}$ and $\frac{9}{4}$

inverse operations— opposite operations, such as multiplication and division

like fractions—fractions that have the same denominator; for example, $\frac{1}{4}$ and $\frac{3}{4}$

numerator—the top number in a fraction; it shows the number of parts used

reciprocal—the result of inverting the numbers in a fraction; for example, the reciprocal of $\frac{2}{3}$ is $\frac{3}{2}$

unlike fractions—fractions with different denominators; for example, $\frac{1}{4}$ and $\frac{1}{3}$

"Fractions aren't just parts of numbers. They are numbers themselves."

How Fractions Work

As you saw in the program, a **fraction** is a way of representing part
of a whole. Whenever we break something whole into equal parts,
we can use a fraction to describe one or more of those parts. The
circle has been divided into four equal parts. One of the four parts
is shaded. In other words, $\frac{1}{4}$ *(one-fourth)* of the circle is shaded.

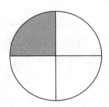

However, fractions are not limited to cutting pies and pizzas into slices. You can also use
a fraction to represent part of a group, a measurement, or a point on a ruler.

At a company, $\frac{3}{4}$ of the managers are women.

The recipe calls for $\frac{1}{3}$ cup of vegetable oil.

The arrow is pointing at $\frac{1}{2}$ on the ruler.

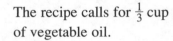
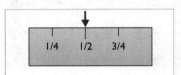

It's important to understand the meaning of the two parts of a fraction. The bottom
number, called the **denominator,** tells how many parts the whole is divided into.
The top number, called the **numerator,** tells how many parts of the whole are used.

$\frac{\text{numerator}}{\text{denominator}}$ $\frac{3}{10}$ $\frac{\text{3 parts are shaded}}{\text{there are 10 equal parts in the object}}$

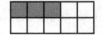

Remember, fractions and decimals can both show an amount less than 1. For example,
the fraction $\frac{3}{10}$ can also be written as a decimal: 0.3. Since decimals use place value,
they always divide the whole into a power of ten: 10, 100, 1,000, and so on. Fractions
do not use place value, so the whole can be divided into any number of equal parts.

Study the number lines below to better understand the relationship between fractions
and decimals.

This number line shows the location of $\frac{4}{5}$.
Find the same location on the decimal
number line. The fraction $\frac{4}{5}$ is equal to $\frac{8}{10}$,
which equals the decimal 0.8.

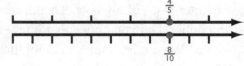

CALCULATOR TIP: If you need to find an equal decimal for any fraction, enter
the fraction as a division problem. To find the decimal that equals $\frac{4}{5}$, try pressing the
following keys: 4 $\boxed{\div}$ 5 $\boxed{=}$. The display reads $\boxed{0.8}$.

PROGRAM 23 ▪ FRACTIONS

Simplifying Fractions

In the last example, you saw that $0.8 = \frac{4}{5}$. Remember that 0.8 is eight-tenths, or $\frac{8}{10}$. That means that $0.8 = \frac{8}{10} = \frac{4}{5}$. Fractions with the same value are called **equivalent fractions**. **Simplifying** a fraction (sometimes called **reducing to lowest terms**) means finding an equivalent fraction with the smallest possible numerator and denominator. To simplify a fraction, divide both the numerator and the denominator by the same number.

E X A M P L E : Write the fraction $\frac{8}{12}$ in simplest terms. Try to think of the greatest number that will divide evenly into 8 and 12. That number is 4. $\quad \frac{8 \div 4}{12 \div 4} = \frac{2}{3}$

Answer: $\frac{8}{12} = \frac{2}{3}$

If you divide by a smaller number first, you can still simplify the fraction, but you will use more steps: $\frac{8 \div 2}{12 \div 2} = \frac{4}{6}$ and $\frac{4 \div 2}{6 \div 2} = \frac{2}{3}$.

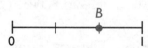

FRACTIONS ▪ PRACTICE I

A. Write a fraction to represent the shaded portion of each figure.

1.
2.
3.

B. Write fractions to represent the points on the number lines.

4.
5.

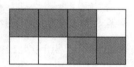

C. Simplify each of these fractions.

6. $\frac{3}{6}$ 8. $\frac{8}{24}$ 10. $\frac{6}{27}$

7. $\frac{12}{20}$ 9. $\frac{30}{40}$ 11. $\frac{10}{25}$

D. Solve. Simplify your answer.

12. In a recent survey, 75 of 125 people said that they disliked surveys. What fraction disliked surveys?

13. Lance earned $120 for a job. His boss deducted $30 for taxes. What fraction of his pay was deducted for taxes?

Answers and explanations start on page 205.
For more practice simplifying and comparing fractions, see page 172.

Add and Subtract Like Fractions

One of these problems doesn't make sense. Which one is it?

1. 3 apples + 4 apples = ?   **2.** 5 apples + 2 inches = ?

The second problem can't be solved. You cannot add 5 apples and 2 inches. This example makes an important point: You can only add or subtract things with a common label. This rule applies to fractions.

In a fraction, the denominator is a label. The fraction $\frac{3}{8}$ means 3 *eighths*. The numerator 3 tells you *how many,* and the denominator *eighths* is the label of the fraction. Two fractions with the same denominator are called **like fractions.** You can only add or subtract like fractions.

E X A M P L E : Add: $\frac{3}{7} + \frac{2}{7}$

Step 1.   Add the numerators: $3 + 2 = 5$.
Step 2.   Write the sum over the shared denominator: $\frac{5}{7}$.

Answer: The sum of $\frac{3}{7}$ and $\frac{2}{7}$ is $\mathbf{\frac{5}{7}}$.

E X A M P L E : Subtract: $\frac{11}{12} - \frac{5}{12}$

Step 1.   Subtract the numerators: $11 - 5 = 6$.
Step 2.   Write the difference over the denominator: $\frac{6}{12}$.
Step 3.   Always simplify your answer: $\frac{6 \div 6}{12 \div 6} = \frac{1}{2}$.

Answer: The difference between $\frac{11}{12}$ and $\frac{5}{12}$ is $\mathbf{\frac{1}{2}}$.

Improper Fractions and Mixed Numbers

When a fraction has the same number on the top and the bottom, its value is 1. The diagram below shows the addition: $\frac{1}{4} + \frac{3}{4} = \frac{4}{4} = 1$, or one whole.

Sometimes the sum of two fractions will be greater than 1; for example, $\frac{7}{8} + \frac{6}{8} = \frac{13}{8}$. The fractions $\frac{4}{4}$ and $\frac{13}{8}$ are called improper fractions. An **improper fraction** has a numerator that is greater than or equal to the denominator.

When the answer is an improper fraction, you should change it to either a whole number or a mixed number. A **mixed number** has a whole number and a fraction part.

EXAMPLE: Change $\frac{13}{8}$ to a mixed number.

Step 1. Divide the numerator by the denominator: $13 \div 8 = 1 \ r5$. The whole number part of the mixed number is 1.

Step 2. To find the fraction part, write the remainder over the denominator of the original fraction: $\frac{5}{8}$. If necessary, simplify the fraction.

Answer: The improper fraction $\frac{13}{8}$ equals the mixed number $1\frac{5}{8}$.

A. Change each fraction to a mixed number, and simplify.

1. $\frac{15}{10}$ 3. $\frac{14}{7}$ 5. $\frac{15}{6}$ 7. $\frac{34}{4}$

2. $\frac{20}{9}$ 4. $\frac{21}{2}$ 6. $\frac{24}{8}$ 8. $\frac{57}{12}$

B. Add or subtract as directed. Simplify your answers.

9. $\frac{3}{8} + \frac{2}{8}$

10. $\frac{3}{4} + \frac{2}{4}$

11. $\frac{10}{11} - \frac{6}{11}$

12. $\frac{14}{15} - \frac{9}{15}$

13. $\frac{5}{6} + \frac{3}{6} + \frac{2}{6}$

14. $\frac{4}{5} + \frac{2}{5} + \frac{4}{5}$

15. $\frac{18}{20} - \frac{12}{20}$

16. $\frac{12}{14} - \frac{8}{14}$

C. First decide whether to add or subtract. Then solve. Simplify your answers.

17. Casey painted the trim in several rooms. She used $\frac{3}{10}$ gallon of paint in the living room, $\frac{4}{10}$ in the bedrooms, and $\frac{5}{10}$ in the kitchen. How many gallons of paint did she use?

18. Mike ran $\frac{7}{8}$ mile on Saturday. He ran $\frac{3}{8}$ mile on Sunday. How much farther did he run on Saturday than on Sunday?

19. The weights of two packages are shown below. What is their total weight?

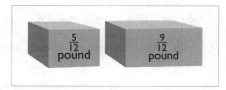

$\frac{5}{12}$ pound $\frac{9}{12}$ pound

20. Simon owns a piece of land that measures $\frac{15}{16}$ acre. If he sells $\frac{9}{16}$ acre, how much land will he have left?

Answers and explanations start on page 205.
For more practice adding and subtracting fractions, see page 173.

"They say you can't add apples and oranges, but you can if you call them both fruit."

Adding and Subtracting Unlike Fractions

Find a Common Denominator

Unlike fractions have different denominators. To add or subtract them, you must first rewrite them as like fractions with a common name, or **common denominator.**

You can rewrite a fraction with a new denominator by **raising the fraction to higher terms.** To raise a fraction, multiply the numerator and the denominator by the same number. This process is the opposite of reducing, or simplifying, a fraction.

E X A M P L E : Raise $\frac{3}{4}$ to a fraction with a denominator of 12.

To raise 4 to 12, you need to multiply by 3. Multiply both the numerator and the denominator by 3.

$$\frac{3 \times 3}{4 \times 3} = \frac{9}{12}$$

Answer: $\frac{3}{4} = \frac{9}{12}$

As you saw in the program, a common denominator for <u>two</u> fractions must be a multiple of <u>both</u> denominators. Remember, multiples are the numbers you say when you count by a number. For example, the multiples of 3 are 3, 6, 9, 12, 15, and so on. The best common denominator to use is the lowest multiple that both numbers share.

E X A M P L E : Find the lowest common denominator for the fractions $\frac{1}{4}$ and $\frac{5}{6}$.

Write the first few multiples of 4: 4, 8, **12,** 16, 20, . . .
Write the first few multiples of 6: 6, **12,** 18, 24, . . .

Answer: The lowest multiple that both numbers have in common is **12.**

Use the lowest common denominator to raise both fractions: $\frac{1 \times 3}{4 \times 3} = \frac{3}{12}$ and $\frac{5 \times 2}{6 \times 2} = \frac{10}{12}$.

You can now add and subtract unlike fractions. Study the following examples. Remember, first find the lowest common denominator. Then raise the fractions to higher terms using the common denominator. Finally, add or subtract.

E X A M P L E : Add: $\frac{2}{3} + \frac{7}{9}$

Step 1. Find a common denominator. If one denominator is a multiple of the other, use the higher number as the common denominator. Since 9 is a multiple of 3, the lowest common denominator is 9.

Step 2. Raise $\frac{2}{3}$ to a fraction with a denominator of 9: $\frac{2 \times 3}{3 \times 3} = \frac{6}{9}$.

Step 3. Add: $\frac{6}{9} + \frac{7}{9} = \frac{13}{9} = 1\frac{4}{9}$.

E X A M P L E : Subtract: $\frac{3}{8} - \frac{1}{6}$

Step 1. Find a common denominator. Write multiples of 6 and 8.

Multiples of 6: 6, 12, 18, **24,** 30, . . .

Multiples of 8: 8, 16, **24,** 32, . . .

The lowest common denominator is 24.

Step 2. Raise the fractions to higher terms.

$$\frac{3}{8} = \frac{3 \times 3}{8 \times 3} = \frac{9}{24} \qquad \frac{1}{6} = \frac{1 \times 4}{6 \times 4} = \frac{4}{24}$$

Step 3. Subtract the like fractions: $\frac{9}{24} - \frac{4}{24} = \frac{5}{24}$.

FRACTIONS ▪ PRACTICE 3

A. Add or subtract as directed. Simplify your answers.

1. $\frac{1}{6} + \frac{2}{3}$

2. $\frac{2}{7} + \frac{3}{14}$

3. $\frac{1}{8} + \frac{5}{6}$

4. $\frac{2}{3} + \frac{8}{15}$

5. $\frac{2}{3} + \frac{3}{4}$

6. $\frac{4}{5} + \frac{2}{7}$

7. $\frac{1}{3} + \frac{1}{4} + \frac{1}{2}$

8. $\frac{2}{5} + \frac{7}{20} + \frac{9}{10}$

9. $\frac{11}{12} - \frac{1}{2}$

10. $\frac{7}{15} - \frac{2}{5}$

11. $\frac{17}{20} - \frac{1}{4}$

12. $\frac{5}{8} - \frac{5}{16}$

13. $\frac{7}{12} - \frac{3}{8}$

14. $\frac{5}{9} - \frac{1}{4}$

15. $\frac{9}{10} - \frac{3}{15}$

B. First decide whether to add or subtract. Then solve. Simplify the answer.

16. The diameters of two bolts are $\frac{13}{16}$ inch and $\frac{1}{2}$ inch. What is the difference in the diameters of the bolts?

17. Joanna had $\frac{5}{6}$ yard of ribbon. If she cuts off a piece that is $\frac{5}{8}$ yard long, what fraction of a yard will be left?

18. The liquid ingredients in a recipe are:

milk	$\frac{3}{4}$ cup
cooking oil	$\frac{1}{3}$ cup
apple juice	$\frac{1}{2}$ cup

What is the total amount of liquid in the recipe?

Answers and explanations start on page 206.

For more practice adding and subtracting fractions, see page 173.

Add and Subtract Mixed Numbers

A **mixed number** has a whole number and a fraction part. When you add mixed numbers, you find the sum of the whole numbers and fractions separately and then combine the two sums.

EXAMPLE: Seth worked $3\frac{1}{2}$ hours on Monday and $5\frac{3}{4}$ hours on Tuesday. How many hours did he work in all?

Step 1. Write the amounts in a column. Rewrite the fractions with like denominators.

Step 2. Add the whole numbers and fractions separately.

$$3\frac{1}{2} = 3\frac{2}{4}$$
$$+\ 5\frac{3}{4} = 5\frac{3}{4}$$
$$8\frac{5}{4} = 8 + 1\frac{1}{4} = 9\frac{1}{4}$$

Step 3. Change the sum of the fractions to a mixed number: $\frac{5}{4} = 1\frac{1}{4}$. Add this sum to the sum of the whole numbers.

Answer: Seth worked **$9\frac{1}{4}$ hours.**

Subtracting a mixed number often involves regrouping. Study the example below to see how to subtract a mixed number from a whole number.

EXAMPLE: Sandra measured $3\frac{1}{4}$ cups of flour from a 10-cup container. How many cups of flour remain in the container?

Step 1. Write the amounts in a column.

Step 2. Rewrite the whole number by regrouping. Borrow 1 from the whole number column and write it as a fraction with the same denominator as the fraction you are subtracting.

$$10 \quad = 9\frac{4}{4}$$
$$-\ 3\frac{1}{4} = 3\frac{1}{4}$$
$$6\frac{3}{4}$$

Step 3. Subtract the fractions and whole numbers separately. Simplify if needed.

Answer: There are **$6\frac{3}{4}$ cups** of flour left in the container.

Study this example to see how to subtract one mixed number from another.

EXAMPLE: $5\frac{1}{3} - 1\frac{1}{2}$

Step 1. Write the problem in a column.

Step 2. Find a common denominator and raise the fractions.

$$5\frac{1}{3} = 5\frac{2}{6} = 4\frac{6}{6} + \frac{2}{6} = 4\frac{8}{6}$$
$$-\ 1\frac{1}{2} = 1\frac{3}{6} = \qquad\qquad 1\frac{3}{6}$$
$$3\frac{5}{6}$$

Step 3. Regroup so that the numerator of the top fraction is greater than the numerator of the bottom fraction.

Step 4. Subtract the fractions and whole numbers separately. Simplify if needed.

Answer: The difference is **$3\frac{5}{6}$.**

A. Solve.

1. $\begin{array}{r} 1\frac{1}{3} \\ + 4\frac{2}{3} \\ \hline \end{array}$

5. $\begin{array}{r} 7\frac{4}{9} \\ + 3\frac{1}{2} \\ \hline \end{array}$

9. $\begin{array}{r} 7\frac{7}{8} \\ - 2\frac{3}{8} \\ \hline \end{array}$

13. $\begin{array}{r} 7\frac{5}{12} \\ - 5\frac{7}{8} \\ \hline \end{array}$

2. $\begin{array}{r} 6\frac{1}{12} \\ + 2\frac{7}{12} \\ \hline \end{array}$

6. $\begin{array}{r} 4\frac{5}{6} \\ + 1\frac{2}{5} \\ \hline \end{array}$

10. $\begin{array}{r} 9\frac{9}{10} \\ - 5\frac{3}{5} \\ \hline \end{array}$

14. $\begin{array}{r} 1\frac{1}{2} \\ - \frac{2}{3} \\ \hline \end{array}$

3. $\begin{array}{r} 4\frac{2}{3} \\ + 8\frac{5}{6} \\ \hline \end{array}$

7. $\begin{array}{r} 4\frac{3}{4} \\ + 2\frac{3}{10} \\ \hline \end{array}$

11. $\begin{array}{r} 12 \\ - 4\frac{2}{7} \\ \hline \end{array}$

15. $\begin{array}{r} 6\frac{1}{7} \\ - 1\frac{5}{21} \\ \hline \end{array}$

4. $\begin{array}{r} 10\frac{7}{10} \\ + 5\frac{4}{5} \\ \hline \end{array}$

8. $\begin{array}{r} 5\frac{1}{3} \\ + 8\frac{5}{12} \\ \hline \end{array}$

12. $\begin{array}{r} 3 \\ - 1\frac{7}{10} \\ \hline \end{array}$

16. $\begin{array}{r} 10\frac{1}{6} \\ - 3\frac{1}{3} \\ \hline \end{array}$

B. First decide whether to add or subtract. Then solve. Simplify the answer.

17. Kelli wants to get a credit line with an interest rate of $7\frac{3}{4}$ percentage points. One bank offers her a rate of $8\frac{9}{10}$ percentage points. How many points higher is this rate than the rate Kelli wants?

18. Jenna is putting molding around a doorway. She needs two lengths measuring $6\frac{3}{4}$ feet and one length measuring $2\frac{3}{4}$ feet. How many total feet of molding does she need?

19. A 9-foot fence post is set $1\frac{2}{3}$ feet into the ground. How much of the post will be above ground?

20. If you cut lengths of $5\frac{5}{8}$ feet and $4\frac{1}{2}$ feet from a 30-foot roll of wallpaper, how many feet of wallpaper will be left on the roll?

21. Scott can work a maximum of 20 hours a week. He has already worked the hours shown below.

Days	Hours
Monday	$2\frac{1}{2}$ hours
Tuesday	4 hours
Wednesday	$3\frac{3}{4}$ hours

How many more hours can he work this week?

Answers and explanations start on page 206.
For more practice adding and subtracting mixed numbers, see page 173.

> *"Don't get so involved in the methods of calculating that you lose track of what fractions mean."*

Multiplying and Dividing Fractions

Multiplication

When you multiply fractions, you don't have to find a common denominator. Just multiply the numerators, then multiply the denominators, and simplify your answer.

EXAMPLE: Multiply: $\frac{2}{3} \times \frac{3}{4}$

Step 1. Multiply the numerators and the denominators.
Step 2. Simplify the answer.

$$\frac{2}{3} \times \frac{3}{4} = \frac{2 \times 3}{3 \times 4} = \frac{6}{12} = \frac{1}{2}$$

As you can see, multiplying fractions is easy, but a simple shortcut can make it even easier. You can use **canceling** to simplify the numbers (and answer) before you multiply. To cancel, divide a numerator and a denominator in the problem by the same number *before* you multiply.

EXAMPLE: Multiply: $\frac{8}{15} \times \frac{3}{5}$

Step 1. Cancel by dividing the 3 (a numerator) and the 15 (a denominator) by 3. Cross out the 3 and the 15, and write the results of the division next to the numbers.

Step 2. Multiply using the new numerators and denominators. The result is in simplest form.

$$\frac{8}{\overset{}{\underset{5}{\cancel{15}}}} \times \frac{\overset{1}{\cancel{3}}}{5} = \frac{8 \times 1}{5 \times 5} = \frac{8}{25}$$

Many problems involve finding a fraction of a whole number or a mixed number. To multiply a fraction by a whole number or a mixed number, write the whole number or mixed number as an improper fraction.

EXAMPLE: Find $\frac{3}{8}$ of 24. (*Hint:* To find an amount "of" another amount, multiply.)

Step 1. Change the whole number to a fraction by writing it over a denominator of 1: $24 = \frac{24}{1}$.
Step 2. Set up the multiplication problem.
Step 3. Cancel. Divide both 24 and 8 by 8.
Step 4. Multiply.

$$\frac{24}{1} \times \frac{3}{8} = \frac{\overset{3}{\cancel{24}}}{1} \times \frac{3}{\underset{1}{\cancel{8}}} = \frac{9}{1} = 9$$

Does 9 seem like a reasonable answer to the last problem? You know that $\frac{3}{8}$ is a little less than $\frac{4}{8}$, or $\frac{1}{2}$. One-half of 24 is 12. The answer should be a little less than 12. Yes, 9 is a reasonable answer.

EXAMPLE: Find $2\frac{1}{2}$ of $6\frac{1}{4}$.

Step 1. Set up the multiplication problem.
Step 2. Change both numbers to improper fractions.
Step 3. There is nothing to cancel, so multiply the numerators and the denominators.
Step 4. Convert the improper fraction to a mixed number.

$$2\frac{1}{2} \times 6\frac{1}{4} = \frac{5}{2} \times \frac{25}{4} = \frac{125}{8} = 15\frac{5}{8}$$

FRACTIONS ▪ PRACTICE 5

A. Multiply.

1. $\frac{7}{8} \times \frac{1}{2}$

2. $\frac{9}{10} \times \frac{2}{5}$

3. $\frac{8}{9} \times \frac{3}{4}$

4. $\frac{1}{6} \times \frac{3}{5}$

5. $\frac{5}{8} \times \frac{1}{10}$

6. $6 \times \frac{3}{4}$

7. $100 \times \frac{4}{5}$

8. $12 \times \frac{1}{6}$

9. $15 \times \frac{2}{3}$

10. $14 \times \frac{7}{8}$

11. $6\frac{1}{2} \times \frac{4}{5}$

12. $10\frac{1}{3} \times 1\frac{1}{2}$

13. $1\frac{7}{8} \times 2\frac{2}{3}$

14. $8\frac{3}{4} \times 1\frac{2}{7}$

15. $4\frac{4}{5} \times 2\frac{1}{12}$

B. Solve. Simplify the answers.

16. A family spends $\frac{2}{5}$ of its take-home pay on rent and utilities. Of this fraction, $\frac{1}{4}$ is spent on utilities. What fraction of its take-home pay is spent on utilities? (*Hint:* Multiply to find a fraction of a fraction.)

17. If you are paid $112 for a full day's work, how much will you be paid if you work $\frac{3}{4}$ of a day?

18. A car drove 70 miles an hour for $2\frac{1}{2}$ hours. How far did the car drive in this time?

Question 19 refers to the map below.

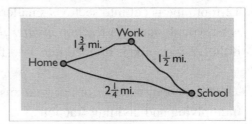

19. Five days a week, Alicia drives the route shown on the map. Starting at home, she drives to work and school and back home. How many miles does she drive per week?

20. Max bought four light fixtures, each with a weight of $1\frac{4}{5}$ pounds. What is the total weight of his order?

Answers and explanations start on page 206.
For more practice multiplying fractions, see page 174.

MATH SKILLS

Division

Multiplication and division have a special relationship. The two operations are opposites, or **inverse operations.** That means that one operation can undo the other operation; for example, $12 \div 2 = 6$ and $6 \times 2 = 12$.

Let's take it a step further. The number 2 can be written as the improper fraction $\frac{2}{1}$. Compare $\frac{1}{2}$ and $\frac{2}{1}$. What do you notice? The fraction $\frac{2}{1}$ is the fraction $\frac{1}{2}$ inverted, or turned upside down. These numbers, called **reciprocals,** are opposites or inverses. If you multiply reciprocals, the product will always be 1: $\frac{1}{2} \times \frac{2}{1} = \frac{2}{2} = 1$.

You can use the relationship between multiplication and division and your knowledge of reciprocals to turn any division problem into a multiplication problem. When you are asked to divide by a fraction, invert the fraction and multiply.

EXAMPLE: A piece of plastic pipe is 12 feet long. How many pieces, each $\frac{3}{4}$ foot long, can be cut from the pipe? (*Hint:* Always write the amount being divided first.)

Step 1.   Write the division problem.
Step 2.   Change to multiplication by inverting $12 \div \frac{3}{4} = 12 \times \frac{4}{3} = \frac{\overset{4}{\cancel{12}}}{1} \times \frac{4}{\cancel{3}_{1}} = 16$
the fraction you are dividing by.
Step 3.   Multiply.

Answer: There are **16 pieces** measuring $\frac{3}{4}$ foot in a 12-foot pipe.

Are you surprised that you ended up with a number larger than 12? When you divide by a fraction *less than 1,* the answer will be greater than the number you divided. When the 12-foot pipe is cut into sections smaller than 1 foot, the result is 16 pieces. However, when you divide by a fraction *greater than 1,* the answer will be smaller. For example, if the 12-foot pipe is divided into $1\frac{1}{2}$ - foot sections, there would be only 8 pieces.

Thinking about whether or not your answer makes sense will greatly help you when multiplying and dividing fractions.

You can divide mixed numbers the same way as you divide fractions. Just change the mixed numbers to improper fractions first.

EXAMPLE: Divide: $3\frac{3}{8} \div 1\frac{1}{2}$

Step 1.   Write the division problem. $3\frac{3}{8} \div 1\frac{1}{2} = \frac{27}{8} \div \frac{3}{2} = \frac{27}{8} \times \frac{2}{3}$
Step 2.   Change the mixed numbers to
improper fractions. $= \frac{\overset{9}{\cancel{27}}}{\underset{4}{\cancel{8}}} \times \frac{\overset{1}{\cancel{2}}}{\underset{1}{\cancel{3}}} = \frac{9}{4} = 2\frac{1}{4}$
Step 3.   Invert the fraction you are dividing
by, and change the operation sign
to multiplication.
Step 4.   Multiply.

Answer: The quotient is $2\frac{1}{4}$.

A. Divide.

1. $\frac{1}{2} \div \frac{2}{3}$

2. $\frac{5}{12} \div \frac{1}{6}$

3. $\frac{4}{9} \div \frac{2}{3}$

4. $\frac{11}{18} \div \frac{1}{9}$

5. $\frac{3}{8} \div \frac{9}{16}$

6. $5 \div \frac{1}{5}$

7. $12 \div \frac{2}{3}$

8. $2\frac{2}{5} \div 12$

9. $1\frac{1}{7} \div 4$

10. $8 \div 3\frac{1}{5}$

11. $2\frac{1}{3} \div 1\frac{5}{9}$

12. $5\frac{1}{4} \div 1\frac{3}{4}$

13. $4\frac{1}{2} \div 6\frac{3}{4}$

14. $2\frac{7}{8} \div \frac{1}{2}$

15. $3\frac{1}{3} \div 1\frac{1}{5}$

B. Write the amount being divided first. Solve. Simplify your answers.

16. A pattern for a child's dress requires $2\frac{1}{8}$ yards of fabric. How many dresses can a tailor make from 34 yards of fabric?

17. A land developer has 21 acres of land to build a housing division. She plans to divide the land into lots measuring $\frac{3}{8}$ acre. How many lots can she make?

18. Patricia drove 175 miles in $3\frac{1}{2}$ hours. She wants to figure out her average speed for the trip. How many miles did the car drive per hour? (*Hint:* Divide the distance she drove by the time in hours. The quotient is her speed in miles per hour.)

19. A stack of books is 35 inches high. If each book in the stack is $1\frac{1}{4}$ inches thick, how many books must be in the stack?

20. How many $\frac{1}{8}$ cups of sugar could you measure from $5\frac{3}{4}$ cups of sugar?

Questions 21 and 22 refer to the table below.

MicroQuest Software Average Length of Customer Calls	
Reason for Call	**Time**
Troubleshooting	$\frac{3}{4}$ hour
Request for Refund	$\frac{1}{3}$ hour

21. Marcia works in technical support for MicroQuest. On average, how many troubleshooting calls can she complete in six hours?

22. Bill works in customer service for MicroQuest. On average, how many requests for refunds can Bill process if he works for $6\frac{2}{3}$ hours?

Answers and explanations start on page 207.
For more practice dividing fractions, see page 174.

The Information You Need

In the video, you saw that fractions are used to solve many everyday problems. For some problems, you will have to select the information you need from other details.

To find **the information you need,** read each problem carefully. Some problems may have more information than you need, while others will not have enough information. For still others, you may need to calculate a needed number.

Too Much Information

In some problems, more numbers are given than you need. In those cases, decide which numbers you need to solve the problem.

> **EXAMPLE:** Leslie ordered 15 boards, each measuring 14 feet long, for framing walls in her storage building. How much of each board will be left after cutting two $5\frac{1}{4}$-foot lengths?
>
> **(1)** $2\frac{3}{4}$ **(4)** $4\frac{1}{2}$
>
> **(2)** $3\frac{1}{4}$ **(5)** $5\frac{1}{4}$
>
> **(3)** $3\frac{1}{2}$

The correct answer is **(3)** $3\frac{1}{2}$. In this problem, you are given extra information. Since you are asked "How much of each board?" the number of boards ordered (15) is not used to solve this problem. Multiply: $2 \times 5\frac{1}{4} = 10\frac{1}{2}$. Subtract: $14 - 10\frac{1}{2} = 3\frac{1}{2}$.

Not Enough Information

In some problems, a piece of information needed to solve the problem is missing.

> **EXAMPLE:** Last week, Paul worked $7\frac{1}{4}$ hours on Monday and Tuesday, $6\frac{1}{2}$ hours on Wednesday, and 8 hours on Thursday and Friday. How many more hours did Paul work this week than last week?
>
> **(1)** 37
>
> **(2)** $21\frac{3}{4}$
>
> **(3)** $8\frac{1}{4}$
>
> **(4)** 3
>
> **(5)** Not enough information is given.

The correct answer is **(5) Not enough information is given.** To answer the question, you need to know how many hours Paul worked last week. None of the information in the problem is about last week, so you cannot solve the problem.

Calculating Needed Information

Some questions may seem to give too little information to solve the problem. However, you may be able to calculate a number from the information given in the problem and then use that number to find the solution.

> **EXAMPLE:** In Marshall's science class, there are 12 boys and 16 girls. What fraction of his students are girls?
>
> **(1)** $\frac{4}{7}$
>
> **(2)** $\frac{3}{4}$
>
> **(3)** $\frac{4}{3}$
>
> **(4)** $\frac{7}{4}$
>
> **(5)** Not enough information is given.

The correct answer is **(1)** $\frac{4}{7}$. In this problem, you need to know the part (how many students are girls) and the whole (the number of students in the class). The part, 16 girls, is given. To find the whole, the total number of students, add $12 + 16 = 28$. The fraction is $\frac{16 \text{ girls}}{28 \text{ total students}}$ or $\frac{16}{28}$. Reduce the fraction to lowest terms: $\frac{16}{28} = \frac{4}{7}$.

THE INFORMATION YOU NEED

Analyze, but <u>do not solve</u> these problems. For each problem, write either *too much, not enough,* **or** *compute needed information* **to show how much information is included. For any problems where you wrote** *not enough,* **write what you would need to solve it.**

1. Joan owes $450 for rent, $35.95 for electricity, and $22.40 for water. How much more will she pay for rent than for the electricity and water combined?

2. Quentin and Tamara invited 50 people to their wedding and reception. If 40 people were at the wedding, how many more people went to the reception than to the wedding?

3. In Smithville, 1,500 people voted Republican and 2,000 voted Democrat in the last election. What fraction of the people voted Democrat?

Questions 4 and 5 refer to the drawing below.

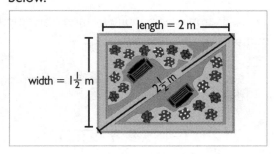

4. What is the total distance around the rose garden?

5. How many rosebushes can be planted in the garden?

Answers and explanations start on page 207.

Fractions on the Calculator and Grid

Not all calculators allow you to use fractions, but the calculator you will use on the GED does have a fraction key. This calculator can be used to add, subtract, multiply, and divide fractions. It is important to check the display on the calculator as you enter a fraction to be sure you enter it correctly.

Calculator Operations

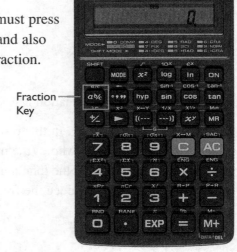

Fraction Key

When entering fractions on the Casio *fx*-260, you must press the fraction key $\boxed{a^{b_{/c}}}$ after any whole number part and also between the numerator and the denominator of a fraction.

For example, to enter $5\frac{1}{2}$:

Enter 5 and press the fraction key: $\boxed{a^{b_{/c}}}$
The display shows: | 5⌐. |

Enter 1 and press the fraction key: $\boxed{a^{b_{/c}}}$
The display shows: | 5⌐1. |

Enter 2.
The display shows: | 5⌐1⌐2. |

The display should show the symbol ⌐ between each part of the fraction and between the whole number and a fraction. If you make a mistake, clear the display and enter the numbers again.

EXAMPLES: Enter each fraction on your calculator. Compare your display with the one shown.

$3\frac{2}{5}$ | 3⌐2⌐5. | $\frac{6}{7}$ | 6⌐7. | $12\frac{5}{8}$ | 12⌐5⌐8. |

These examples show one or more operations with fractions, whole numbers, and mixed numbers.

EXAMPLES: Enter each problem on your calculator. Compare your answer with the display.

$\frac{3}{10}$ $\boxed{+}$ $\frac{4}{5}$ $\boxed{=}$ | 1⌐1⌐10. | Write the answer as $1\frac{1}{10}$.

$4\frac{1}{4}$ $\boxed{-}$ $2\frac{1}{8}$ $\boxed{=}$ | 2⌐1⌐8. | Write the answer as $2\frac{1}{8}$.

$\frac{2}{3}$ $\boxed{\times}$ 7 $\boxed{=}$ | 4⌐2⌐3. | Write the answer as $4\frac{2}{3}$.

9 $\boxed{\div}$ $\frac{1}{3}$ $\boxed{=}$ | 27. |

Grid Basics

When filling in a fraction on the standard grid:

- Write the answer in fraction form or change it to decimal form.
- Write any mixed number as a decimal or an improper fraction.
- Write the fraction bar (or decimal point) in its own column.

EXAMPLE: A road sign shows Exit 34 at $3\frac{1}{2}$ miles and Exit 35 at $4\frac{3}{4}$ miles. How far apart are the two exits?

The grids show the answer in both fraction and decimal form.

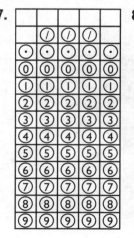

Note: As a fraction, $1\frac{1}{4}$ has to be written as 5/4 in the grid.

GED MATH PRACTICE

FRACTIONS ON THE CALCULATOR AND GRID

Use the fraction key on your calculator to solve each problem below.

1. $5\frac{1}{4} + 2\frac{3}{4} =$

2. $\frac{9}{10} \div \frac{3}{10} =$

3. $8\frac{2}{3} \times \frac{1}{2} =$

4. $\frac{7}{16} - \frac{3}{8} =$

5. $\frac{3}{4} + \frac{1}{8} \times \frac{1}{2} =$

6. $6\frac{4}{5} - 2\frac{1}{10} =$

Mark your answers on the grids provided.

<u>Questions 7 and 8</u> refer to the graph below.

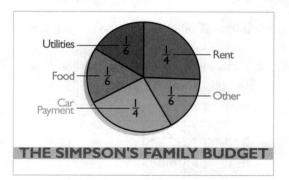

THE SIMPSON'S FAMILY BUDGET

7. What fraction of the total budget goes for rent and utilities?

8. If the total budget is $1,500 per month, how many dollars are spent on car payments?

Answers and explanations start on page 207.
For more practice with fraction operations and grids, see page 175.

GED Review: Fractions

GED REVIEW

Part I: Choose the <u>one best answer</u> to each question below. You <u>may</u> use your calculator.

1. If 1 yard = 3 feet, how many feet are in $2\frac{3}{4}$ yards?
   - **(1)** $5\frac{3}{4}$
   - **(2)** $6\frac{1}{4}$
   - **(3)** $7\frac{1}{3}$
   - **(4)** $7\frac{1}{2}$
   - **(5)** $8\frac{1}{4}$

2. A storage container is $16\frac{1}{2}$ inches long, $11\frac{1}{2}$ inches wide, and 6 inches high. How many more inches is the container long than wide?
   - **(1)** 5
   - **(2)** $10\frac{1}{2}$
   - **(3)** $17\frac{1}{2}$
   - **(4)** 28
   - **(5)** 34

3. On an exercise trail, Kaneesha walked $1\frac{3}{10}$ miles east, $\frac{7}{10}$ mile south, and 1 mile west. If she continues to follow the trail, how many miles is she from where she started?

   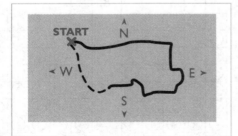

   - **(1)** 1
   - **(2)** $1\frac{1}{10}$
   - **(3)** $1\frac{1}{2}$
   - **(4)** $2\frac{3}{10}$
   - **(5)** Not enough information is given.

4. How many inches are between $23\frac{3}{16}$ inches and $24\frac{1}{16}$ inches on the measuring tape shown below?

   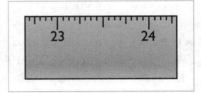

   - **(1)** $\frac{1}{16}$
   - **(2)** $\frac{3}{16}$
   - **(3)** $\frac{3}{8}$
   - **(4)** $\frac{7}{8}$
   - **(5)** $1\frac{1}{8}$

5. Jordon wants to make only $\frac{1}{3}$ of the waffle recipe that calls for 2 cups mix, $1\frac{1}{2}$ cups milk, 1 egg, and 2 tablespoons of oil. How many cups of milk will he use?
   - **(1)** 1
   - **(2)** $\frac{3}{4}$
   - **(3)** $\frac{1}{2}$
   - **(4)** $\frac{1}{4}$
   - **(5)** Not enough information is given.

6. Jamal is building a retaining wall from logs. Each log measures $7\frac{1}{4}$ inches in diameter. How many logs will he need to build a wall 58 inches tall?

Part II: Choose the <u>one best answer</u> to each question below. You <u>may not</u> use your calculator.

<u>Questions 7 and 8</u> refer to the chart below.

Employee	Weekly Hours Worked
Monica	35
Tom	$26\frac{1}{4}$
Wendy	$38\frac{3}{4}$
Frank	40
Total	140

7. What fraction of the total number of hours did Monica work?
   - **(1)** $\frac{3}{16}$
   - **(2)** $\frac{1}{4}$
   - **(3)** $\frac{1}{3}$
   - **(4)** $\frac{5}{8}$
   - **(5)** $\frac{3}{4}$

8. How many more hours did Frank work than Tom?
   - **(1)** $113\frac{3}{4}$
   - **(2)** 100
   - **(3)** 67
   - **(4)** $13\frac{3}{4}$
   - **(5)** $12\frac{1}{2}$

9. A recipe calls for $1\frac{3}{4}$ cups flour. If George triples the recipe, how many cups of flour will he need?
   - **(1)** $3\frac{3}{4}$
   - **(2)** $5\frac{1}{4}$
   - **(3)** 7
   - **(4)** $8\frac{1}{4}$
   - **(5)** $9\frac{1}{2}$

10. Max is recovering his dining room chairs. He needs $\frac{1}{2}$ yard of fabric for each chair seat and $\frac{3}{4}$ yard for each chair back. Which expression can be used to find how many yards of fabric he needs for the 6 chairs?
    - **(1)** $\frac{1}{2} + \frac{3}{4}$
    - **(2)** $\frac{1}{2} + \frac{3}{4} \times 6$
    - **(3)** $(\frac{1}{2} + \frac{3}{4}) \times 6$
    - **(4)** $\frac{1}{2} \times \frac{3}{4} \times 6$
    - **(5)** $\frac{1}{2} \times \frac{3}{4}$

11. Nathan started the week with $50. On Monday he spent $\frac{1}{2}$ of his money on groceries. On Friday he spent $\frac{3}{4}$ of his remaining money on dinner. How much does he have left?
    - **(1)** $25.00
    - **(2)** $18.75
    - **(3)** $12.50
    - **(4)** $6.25
    - **(5)** Not enough information is given.

12. Omar bought $7\frac{1}{2}$ pounds of ground beef for a family cookout. How many $\frac{1}{4}$-pound burgers can he make from the ground beef?

Answers and explanations start on page 207.

Ratio, Proportion, and Percent

LESSON GOALS

MATH SKILLS

- Use ratios and rates
- Set up and solve proportions
- Solve percent problems

GED PROBLEM SOLVING

- Use proportions to solve problems

GED MATH CONNECTION

- Percents on the calculator and the grid

GED REVIEW

EXTRA PRACTICE PP. 140–143

- Ratio and Proportion
- Percents
- Percents with Calculators and Grids

1. Think About the Topic

The program that you are going to watch is about *Ratio, Proportion, and Percent*. The video will show that ratio and proportion are ways of comparing values. You will see what these terms mean and how to work with them. In addition, you will see how percents are used.

2. Prepare to Watch the Video

This program will give ideas about developing your understanding of ratios, proportions, and percents. Answer the questions below to see how much you already know.

If your insurance will pay 100 percent of a $625 bill, how much will your insurance pay?

_____

If the insurance company pays 100 percent of the bill, then that means it will pay the whole bill, or $625.

If a case of soda pop is on sale for $5, how much would 3 cases cost (before tax)?

_____

Since 1 case is $5, 2 cases would be $10, and 3 cases would be $15. You may not know it, but you used proportion to answer this question. Understanding and using proportion will help you pass the GED Math Test.

3. Preview the Questions

Read the questions under *Think About the Program* below, and keep them in mind as you watch the program. You will be reviewing them after you watch.

4. Study the Vocabulary

Review the terms to the right. Understanding the meaning of math vocabulary will help you understand the video and the rest of this lesson.

WATCH THE PROGRAM

As you watch the program, pay special attention to the host who introduces or summarizes major ideas that you need to learn about. The host will also provide important information about the GED Math Test.

AFTER YOU WATCH

1. Think About the Program

What are three ways to write or show ratios?

What is the whole amount that every percent is compared with?

How would you describe a proportion?

What is a common use of ratios?

2. Make the Connection

The program talked about percents and sales. How do you figure in your head how much you will save on sale items? What percents are easy to work with in your head? Why?

TERMS

cross multiply—a process to show two ratios are equal; for example, to see if $\frac{25}{10} = \frac{5}{2}$, multiply $25 \times 2 = 50$ and $10 \times 5 = 50$; the ratios are equal

part—in a percent situation, a portion of the whole

percent of change—a percent ratio where the amount of change is compared with the original amount

percent—a ratio that compares a number with 100; shown by the % symbol

proportion—a math statement that two ratios are equal

rate (definition 1)—also called *unit rate*; a special ratio with a denominator of 1

rate (definition 2)—the element in a percent situation that is followed by the % symbol

ratio—a way of using division to compare two numbers

successive percents—a series of percent calculations where each new calculation depends on the result of the one before it

term—one of the values in a proportion

whole—in a percent situation, the base, or entire, amount

Ratios and Rates

Working with Ratios

A **ratio** is used to compare two numbers. We use ratios to understand the relationship between two numbers. For example, suppose a baseball team has won 6 of the 12 games it has played. One way to make sense of this information is to write a ratio. You could say that the ratio of *games won* to *games played* is 6 to 12.

Ratios can be written in words using the word *to,* with a colon, and as a fraction. Each of the following ratios is read "6 to 12."

$$6 \text{ to } 12 \qquad 6{:}12 \qquad \frac{6}{12}$$

Ratios are more meaningful when they are simplified. To simplify a ratio, write it as a fraction, and reduce it to lowest terms: $\frac{6}{12} = \frac{6 \div 6}{12 \div 6} = \frac{1}{2}$. Using the simplified ratio, you could say that the ratio of games won to games played is 1 to 2. In other words, the team has won 1 game out of every 2 games it has played.

Even though ratios can be written in fraction form, they do not follow all the rules of fractions. If a ratio looks like an improper fraction, do not change it to a mixed number. It is possible for a ratio to have a larger number on the top than on the bottom.

EXAMPLE: An animal shelter finds homes for dogs and cats. Of the animals waiting for adoption, 21 are dogs and 14 are cats. What is the ratio of dogs to cats at the shelter?

When you solve a ratio problem, read the question carefully. To be correct, the numbers in the ratio must be written in the order stated in the question. This problem asks you to write a ratio comparing the number of dogs to the number of cats. Write the ratio and simplify: $\frac{21}{14} = \frac{21 \div 7}{14 \div 7} = \frac{3}{2}$.

Answer: The ratio of dogs to cats is $\frac{3}{2}$, **3:2,** or **3 to 2.**

Some ratio problems have more than one step. You may need to solve for one of the numbers in the ratio before you can write the ratio.

EXAMPLE: Chelsea took a math test with 40 questions. She answered 35 questions correctly. What is the ratio of <u>correct to incorrect answers</u>?

Step 1. The question asks you to compare correct to incorrect answers, but the problem doesn't tell you the number of incorrect answers: 40 questions − 35 correct answers = 5 incorrect answers.

Step 2. Write the ratio. Make sure the numbers are in the order stated in the question. Simplify the ratio.

$$\frac{\text{correct answers}}{\text{incorrect answers}} \quad \frac{35}{5} = \frac{35 \div 5}{5 \div 5} = \frac{7}{1}$$

Answer: The ratio of correct to incorrect answers is $\frac{7}{1}$, **7:1**, or **7 to 1.**

RATIO, PROPORTION, AND PERCENT ■ PRACTICE 1

A. Write each ratio as a fraction, and simplify if necessary.

Example: 8:16 $\frac{8}{16} = \frac{1}{2}$

1. 21:3 _____
2. 12 to 15 _____
3. 4:7 _____

4. 6 to 27 _____
5. 16:9 _____
6. 100 to 25 _____

7. 42:30 _____
8. 25 to 40 _____
9. 11:121 _____

B. Solve. Simplify your answers.

10. Melinda earns $240 per week. Tony earns $180 per week. What is the ratio of Melinda's to Tony's earnings?

11. In a city election, a proposition received 4,500 yes votes and 1,800 no votes. What is the ratio of yes to no votes?

12. Craig worked 25 hours on a construction job and 15 hours on a data-entry job. What is the ratio of the time spent on the construction job to the total time worked on the two jobs?

13. Lindsey's gross pay is $270. Her employer takes out $54 in deductions. What is the ratio of Lindsey's gross pay to her take-home pay? (*Hint:* Take-home pay is equal to gross pay minus deductions.)

Questions 14 through 18 refer to the following table.

Softball League Results		
Team	**Wins**	**Losses**
Kryptonite	21	9
Bandits	18	12
Thunder	15	15
High Heat	6	24

14. What is the ratio of Kryptonite wins to Bandit wins?

15. What is the ratio of Thunder losses to the total number of games the team played?

16. Which team has a win-loss ratio of 7 to 3?

17. What is the ratio of High Heat losses to Bandit losses?

18. Which team has a loss-win ratio of 4 to 1?

Answers and explanations start on page 208.
For more practice with ratios, see page 176.

Working with Rates

Many common situations are actually rates.

- A cyclist travels 40 miles per hour.
- A grocery store charges $3.98 per pound for extra-lean ground beef.
- In the city, a car gets 25 miles per gallon of gasoline.

The word *per* tells you that these situations are rates. A **unit rate** is a special ratio with a denominator of 1. To find the unit rate, write the ratio as a fraction. Then divide the numerator by the denominator so that the fraction has a denominator of 1.

EXAMPLE: Annie drove 408 miles on 12 gallons of gasoline. How many miles per gallon did Annie average on her trip?

Step 1. Write the ratio, comparing miles to gallons: $\frac{408 \text{ miles}}{12 \text{ gallons}}$.

Step 2. Divide: $408 \div 12 = 34$, so $\frac{408 \text{ miles}}{12 \text{ gallons}} = \frac{34 \text{ miles}}{1 \text{ gallon}}$.

Answer: Annie averaged **34 miles per gallon** of gasoline on her trip.

As you saw in the video, comparing unit rates can help you save money when you shop. Many grocery stores put a label on the shelf beneath each product. The label tells the unit rate, or price per measurement unit, for the product. Some common measurement units are ounces, pounds, kilograms, and servings.

When you take math tests such as the GED Math Test, you may be asked to calculate and compare unit rates.

EXAMPLE: Jim is buying dog food. Proud Pet dog food comes in 22-ounce cans and costs $0.77. Pet Gourmet dog food comes in a 15-ounce can and costs $0.57. Which brand is the better buy?

To find the better buy, you need to figure out which brand has the lower unit price (price per ounce of food).

Step 1. Write ratios for each brand, comparing price to ounces.	**Step 2.** Divide to find the unit rate, or price per ounce.	**Step 3.** Compare.
Proud Pet: $\frac{\$0.77}{22 \text{ ounces}}$	$\$0.77 \div 22 = \$0.035$	Unit rate: 3.5 cents per ounce
Pet Gourmet: $\frac{\$0.57}{15 \text{ ounces}}$	$\$0.57 \div 15 = \$0.038$	Unit rate: 3.8 cents per ounce

Answer: Although a can of Pet Gourmet dog food costs less than a can of Proud Pet dog food, the Proud Pet brand costs less per ounce. **Proud Pet** is the better buy.

A. Find each unit rate.

1. 540 calories for 3 servings
2. 24 customers in 4 hours
3. $108 for 9 square yards of carpet
4. 5,280 feet in 12 minutes
5. 32 ounces in 64 bags

6. $90 in 6 hours
7. 60 cents for 5 ounces
8. 224 children for 16 teams
9. 1,200 meters in 8 minutes
10. 220 miles in 4 hours

B. Find the best buy in each advertisement.

11.

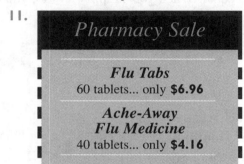

Pharmacy Sale

Flu Tabs
60 tablets... only **$6.96**

**Ache-Away
Flu Medicine**
40 tablets... only **$4.16**

12.

BACK-TO-SCHOOL SALE

FINELINE PENS
(8-pack) $7.12

BESTINK PENS
(12-pack) $10.32

WRITEAWAY PENS
(10-pack) $9.20

C. Solve.

13. Jennifer paid $3.90 for a five-pound bag of potting soil. What was the price per pound of the potting soil?

14. David saved $7,500 in six years. On average, how much did David save per year?

15. According to Kimball's phone bill, he paid $1.12 to make a 16-minute long-distance call. How many cents per minute does Kimball pay for long-distance calls?

16. Mikal has to drive 275 miles to make a delivery. How many miles per hour will he need to average to make the trip in five hours?

17. Hank drove 360 miles on 15 gallons of gasoline. How many miles per gallon did Hank average on his trip?

18. Zach needs laser printer paper. He can buy a box of three reams of the paper for $21.60 or a box of five reams for $36.25. Which box has the best price per ream?

19. Min Lee needs to buy AA batteries. A store has the following special:

    5-pack AA batteries.........$3.90
    8-pack AA batteries.........$5.92

    How many cents less per battery will she pay if she buys the 8-pack?

Answers and explanations start on page 208.
For more practice with rates, see page 176.

"One of the things that we do in the world of visual effects is use proportion to figure out how to make something appear to be bigger than it is."

Proportions

Writing Proportions

A **proportion** is a statement that two ratios are equal. When you simplify a ratio, you are actually writing a proportion.

For instance, a baseball pitcher struck out 25 batters and walked 10. His ratio of *strikeouts* to *walks* is $\frac{25}{10}$. Simplify the ratio by dividing both numbers by 5.

$$\frac{25 \div 5}{10 \div 5} = \frac{5}{2}$$

The statement $\frac{25}{10} = \frac{5}{2}$ means that both ratios are equal. To prove that two ratios are equal, you can **cross multiply.** To cross multiply, multiply the denominator of each ratio by the numerator of the other ratio. Follow the arrows in the diagram. If the products are equal, then this is a proportion.

$$2 \times 25 = 50 \qquad 10 \times 5 = 50$$
$$\frac{25}{10} \diagup\!\!\!\!\diagdown \frac{5}{2}$$

Solving Proportions

In a proportion problem, one of the numbers, or **terms,** is missing. You can use your understanding of fractions and cross multiplication to find the missing number.

EXAMPLE: Allyson is making a dessert that calls for 2 cups of brown sugar and 3 cups of flour. She wants to make more servings. How many cups of flour should she use with 4 cups of brown sugar?

Step 1. Set up the proportion. Write the labels first so that you can easily remember what the numbers represent. Keep the labels on both sides of the ratio in the same order *(sugar to flour)*. In the second ratio, let the letter x represent the unknown number

$$\frac{\text{cups of brown sugar}}{\text{cups of flour}} \quad \frac{2}{3} = \frac{4}{x}$$

Step 2. Cross multiply two known terms: $3 \times 4 = 12$.

Step 3. Divide by the remaining term to solve for x: $12 \div 2 = 6$.

Answer: Allyson should use **6 cups** of flour.

After writing the proportion, you may notice that Allison is doubling the recipe. This problem could be easily solved without cross multiplying and dividing. Setting up a proportion may let you "see" the solution to a problem by raising or reducing a fraction.

Since $\frac{2}{3} = \frac{4}{6}$, you can see that the answer is **6 cups** of flour.

If you can't use raising or reducing to solve the proportion, you must cross multiply and divide. This method will always work to find the missing number. In the next example, the unknown value is the top number in the second ratio. The process is still the same.

E X A M P L E : $\frac{6}{8} = \frac{x}{28}$

Step 1. Cross multiply: $28 \times 6 = 168$.
Step 2. Divide by the third number: $168 \div 8 = 21$.

Answer: The value of x is **21**.

RATIO, PROPORTION, AND PERCENT ■ PRACTICE 3

A. Solve for the missing number in each proportion.

1. $\frac{2}{3} = \frac{10}{x}$

2. $\frac{3}{5} = \frac{x}{20}$

3. $\frac{7}{8} = \frac{35}{x}$

4. $\frac{6}{4} = \frac{x}{9}$

5. $\frac{9}{11} = \frac{27}{x}$

6. $\frac{45}{25} = \frac{x}{10}$

7. $\frac{4}{24} = \frac{6}{x}$

8. $\frac{3}{4} = \frac{x}{28}$

9. $\frac{26}{12} = \frac{13}{x}$

10. $\frac{1.5}{6} = \frac{x}{16}$

11. $\frac{8}{0.5} = \frac{32}{x}$

12. $\frac{5}{25} = \frac{x}{2}$

B. Write a proportion and solve.

13. Kina worked 4 hours and received $33 in pay. At that rate, how much would she be paid for 32 hours of work?

$$\frac{\text{hours}}{\text{dollars}} \quad \frac{4}{\$33} = \underline{}$$

14. A machine can make 25 products in 5 minutes. How many products could the machine make in 60 minutes?

$$\frac{\text{products}}{\text{minutes}} \quad \frac{25}{} = \frac{x}{}$$

15. Robert drove 156 miles on 6 gallons of gasoline. At that rate, how far could he drive on a full tank of 15 gallons?

$$\frac{\text{miles}}{\text{gallons}} \quad \frac{}{6} = \underline{}$$

16. The instructions on a drink mix read:

Drink Mix

Add 2 scoops of drink mix to every 6 cups of water.

How many scoops would you add to 9 cups of water?

$$\frac{\text{scoops}}{\text{cups of water}} \quad \frac{}{} = \frac{}{}$$

Answers and explanations start on page 208.
For more practice with proportion, see page 177.

Proportion Applications

Proportion is a good way to solve many types of problems. Pay close attention to the three kinds of problems in this section.

Rate Problems

Stores often encourage consumers to buy more than one of an item by offering a special rate. For example, a store may advertise, "3 cans for $1." What happens when a customer decides to buy more than three cans? You need proportion to find out.

EXAMPLE: At Mom's Market, two cartons of strawberries cost $1.46. At the same rate, what is the cost of five cartons?

Step 1. Set up the proportion. Make sure both ratios are in the same order.

$$\frac{\text{cartons}}{\text{price}} \quad \frac{2 \text{ cartons}}{\$1.46} = \frac{5 \text{ cartons}}{x}$$

Step 2. Solve for the missing number.

$$\$1.46 \times 5 = \$7.30 \quad \$7.30 \div 2 = \$3.65$$

Answer: The cost of five cartons is **$3.65.**

In the problem, the words *at the same rate* are a clue that a proportion can be used to solve the problem. These words tell you that the ratios described are equal.

Scale Drawings and Maps

Scale drawings and maps also use proportion. A map scale compares the distance on a map to actual distance. For example, a map scale might state that 1 inch = 50 miles. This means that every inch on the map is equal to 50 miles in real distance. Study this example to see how proportion is used to work with map scales.

EXAMPLE: On a map, the distance between Wayne Campground and Greenville is $3\frac{1}{2}$ inches. Using the map scale, how many miles is it from the campground to Greenville?

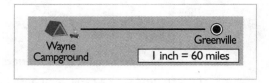

Step 1. Set up the proportion. The terms in a proportion can include fractions. Notice the process for solving the proportion is the same.

$$\frac{\text{inches}}{\text{miles}} \quad \frac{1 \text{ in.}}{60 \text{ mi.}} = \frac{3\frac{1}{2} \text{ in.}}{x \text{ mi.}}$$

Step 2. Solve for the missing number.

$$60 \times 3\frac{1}{2} = 210 \quad 210 \div 1 = 210 \text{ miles}$$

Answer: The actual distance is **210 miles.**

Recipes and Mixtures

Some proportion problems state one of the ratios using a colon. These problems often involve recipes and mixtures.

E X A M P L E : Paint color #A42 calls for mixing blue and white paint at a ratio of 1:8. How many quarts of blue paint should a hardware store clerk add to 12 quarts of white paint?

Step 1. Set up the proportion. The first ratio is the mixture ratio.

$$\frac{\text{blue paint}}{\text{white paint}} \qquad \frac{1}{8} = \frac{x}{12}$$

Step 2. Solve for the missing number.

$$1 \times 12 = 12 \qquad 12 \div 8 = \mathbf{1.5} \text{ or } 1\tfrac{1}{2} \text{ quarts}$$

RATIO, PROPORTION, AND PERCENT ■ PRACTICE 4

Solve.

1. A store ran the following advertisement in the newspaper.

> **DVD Classics**
> 3 titles for only $25.50

At the same rate, how much would a customer pay for ten DVD titles?

<u>Questions 2 and 3</u> are based on the label below.

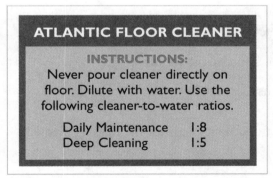

> **ATLANTIC FLOOR CLEANER**
>
> **INSTRUCTIONS:**
> Never pour cleaner directly on floor. Dilute with water. Use the following cleaner-to-water ratios.
>
> Daily Maintenance   1:8
> Deep Cleaning   1:5

2. Jillian wants to fill a spray bottle with floor cleaner for daily maintenance. She plans to put 32 ounces of water in the bottle. How many ounces of floor cleaner should she add to the water?

3. Matt is using Atlantic Floor Cleaner before refinishing. If he uses the deep cleaning ratio, how many quarts of water should he add to $1\tfrac{3}{4}$ quarts of cleaner?

4. A map has a scale of $\tfrac{1}{2}$ inch = 200 miles. What is the distance between two cities that are $2\tfrac{1}{4}$ inches apart on the map?

5. A hardware store is selling 5 rolls of masking tape for $7.00. At the same rate, how much will 18 rolls of tape cost?

<u>Question 6</u> refers to the map below.

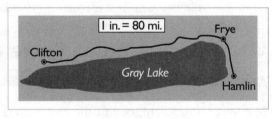

> 1 in. = 80 mi.
> Clifton
> Frye
> Gray Lake
> Hamlin

6. If the distance from Clifton to Frye is 140 miles, what is the distance on the map?

7. An artist mixes white and green oxide paint in a ratio of 7:2. How many ounces of white should the artist add to 6 ounces of green oxide?

8. A sporting goods store advertises basketball jerseys at the price of 4 for $50. At the advertised rate, how many jerseys can a customer buy for $125?

Answers and explanations start on page 208.
For more practice solving proportions, see page 177.

"If you're going to finance a big purchase, you need to understand percentages."

Percents

The Meaning of Percent

A percent is like a fraction or a decimal because it describes part of a whole. The difference is that a **percent** is a ratio that always represents the whole as 100. Percent is shown using the symbol %.

Think about the fraction $\frac{1}{2}$. If you have one-half of something, you have 50% of it. What does 50% really mean? It means 50 out of 100, or $\frac{50}{100}$. In fact, we can write a proportion using these two ratios: $\frac{1}{2} = \frac{50}{100}$.

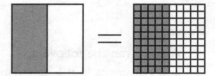

A percent problem is actually a proportion problem. In a percent problem, there are three numbers: the **part,** the **whole,** and the **rate.** The part is always compared to the whole. The rate is always compared to 100.

E X A M P L E : Phillip earns $2,000 per month in take-home pay. He spends 25% of his income, or $500, on rent. Identify the part, the whole, and the rate.

Answer: Whole: $2,000 (Phillip's monthly income)   **Part:** $500 (rent)   **Rate:** 25%
Hint: The rate is the easiest element to find because it is always followed by the % symbol.

Use Proportion to Solve Percent Problems

Any percent situation can be set up as a proportion. Use the format $\frac{part}{whole} = \frac{rate}{100}$.

In a percent problem, one of the elements is missing: the part, the whole, or the rate.
- Decide which element is missing.
- Write the two known elements in the proportion. Use x for the missing element.
- Then solve the proportion.

E X A M P L E : On a test, Andrew correctly answered 46 out of 50 problems. What percent did he answer correctly?

Step 1.   Identify the elements. The whole is 50, the number of test questions. The part is 46, the number Andrew answered correctly. You need to find the percent (rate).

Step 2.   Write a proportion: $\frac{part}{whole} = \frac{rate}{100}$ $\frac{46}{50} = \frac{x}{100}$

Step 3.   Solve: $46 \times 100 = 4,600$, and $4,600 \div 50 = 92$. $\frac{92}{100} = 92\%$

Answer: Andrew answered **92%** of the questions correctly.

E X A M P L E : What is 35% of $60?

Step 1. Identify the elements. The rate is 35%, but is $60 the part or the whole? Remember, in a percent situation, the part is part "of" the whole. The number that follows "of" is the whole. The whole is $60. You need to find the part.

Step 2. Write a proportion: $\frac{\text{part}}{\text{whole}} = \frac{\text{rate}}{100}$ $\frac{x}{60} = \frac{35}{100}$

Step 3. Solve: $60 \times 35 = 2{,}100$ and $2{,}100 \div 100 = 21$.

Answer: 35% of $60 is **$21.**

E X A M P L E : 8 is 20% of what number?

Step 1. Identify the elements. The number that follows "of" is missing, so you need to find the whole. The part is 8, and the rate is 20%.

Step 2. Write a proportion: $\frac{\text{part}}{\text{whole}} = \frac{\text{rate}}{100}$ $\frac{8}{x} = \frac{20}{100}$

Step 3. Solve: $8 \times 100 = 800$, and $800 \div 20 = 40$.

Answer: 8 is 20% of **40.**

RATIO, PROPORTION, AND PERCENT ▪ PRACTICE 5

A. Write a proportion, and solve for the missing element.

1. What number is 70% of 90?
2. $60 is what percent of $500?
3. 15% of what number is 3?
4. Find 75% of $8.

5. What percent of $36 is $9?
6. What is 30% of 80?
7. 19 is 50% of what number?
8. What percent is 22 of 25?

B. Solve.

9. Eighteen out of 20 students in a class have part-time jobs. What percent of the students work part-time?

10. A car dealer has 40 used cars in stock. During a sales promotion, he sells 65% of the cars. How many used cars did he sell?

11. A newspaper ran a survey to see whether people were in favor of a tax hike. Of those surveyed, 132, or 15%, were in favor of the tax increase. How many people were surveyed?

12. At a factory, 21 out of 700 products had faulty switches. What percent of the products had faulty switches?

13. A can of mixed nuts has the following ingredients:

peanuts	6 ounces
cashews	3 ounces
pecans	2 ounces
hazelnuts	1 ounce

What percent of the total nuts are peanuts and cashews?

Answers and explanations start on page 209.
For more practice with percents, see page 178.

MATH SKILLS

Multi-Step Percent Problems

Sales and Discounts

Discounts and sales are common multi-step percent problems.

EXAMPLE: At a golf store, a titanium driver is regularly priced at $380. This week the driver is on sale for 20% off. What is the sale price of the driver?

Step 1. First find the amount of the discount. The regular price ($380) is the whole, the discount is the part, and the rate of the discount is 20%. Find 20% of $380: $\frac{x}{\$380} = \frac{20}{100}$ $\$380 \times 20 = \$7,600$ $\$7,600 \div 100 = \$76$

Step 2. Subtract the discount from the original price: $380 − $76 = **$304.**

Answer: The sale price is **$304.**

Mental Math Shortcut Here is another way to work this problem. You can use mental math to skip a step. Since the discount is 20%, the sale price must be 80% of the regular price. Think of it this way: 100% (the regular price) − 20% (the discount) = 80% (the sale price). Now you can solve for the sale price by finding 80% of $380.

Find 80% of $380: $\frac{x}{\$380} = \frac{80}{100}$ $\$380 \times 80 = \$30,400$ $30,400 \div 100 = $ **$304**

The answer is the same either way. You decide which way seems easier.

Successive Percents

In the next example, a product is discounted 30% and then 20% more. That seems to be a 50% discount overall, but percents don't work that way. When one percent comes after the other, they are called **successive percents.** When percents are successive, the second answer depends on the first answer. You have to work them one at a time.

EXAMPLE: The regular price of a digital camera was $500. When the camera didn't sell, the manager took 30% off the regular price. When it still didn't sell, she reduced the price another 20%. What is the new sale price of the camera?

Step 1. Find the first discount. Find 30% of $500, and subtract:
$\frac{x}{\$500} = \frac{30}{100}$ $\$500 \times 30 = \$15,000$ $\$15,000 \div 100 = \$150$ $\$500 − \$150 = \$350$

Step 2. Find the second discount. Find 20% of $350, and subtract:
$\frac{x}{\$350} = \frac{20}{100}$ $\$350 \times 20 = \$7,000$ $\$7,000 \div 100 = \$70$ $\$350 − \$70 = \$280$

Answer: The new sale price is **$280.**

Percent of Change

MATH SKILLS

E X A M P L E : Ann earned $2,900 per month. After an excellent work review, she was given a raise. Her new monthly salary is $3,132. By what percent did her salary increase?

Ann's raise is the difference between $2,900 and $3,132. To find the percent that her salary increased, you need to find what percent the raise is of $2,900.

Use this proportion to find a percent of change: $\frac{\text{amount of change}}{\text{original amount}} = \frac{\text{rate}}{100}$

Step 1. Subtract to find the amount of change: $3,132 - $2,900 = $232.

Step 2. Use the proportion. The original amount is the amount that came first. In this case, the original amount is Jessica's salary before she received the raise.

$$\frac{\$232}{\$2,900} = \frac{x}{100} \qquad \$232 \times 100 = \$23,200 \qquad \$23,200 \div \$2,900 = 8 \qquad \frac{8}{100} = 8\%$$

Answer: Ann received an **8%** raise.

 RATIO, PROPORTION, AND PERCENT ▪ PRACTICE 6

Solve.

1. Tickets to a new musical are $35 each. Customers can save 20% if they buy their tickets before August 25. What is the cost of a ticket on July 26?

2. Last year a charity raised $12,000. This year it raised $15,000. What was the percent of increase from last year to this year?

3. Courtney went to a one-day sale at a clothes store. A sign in the window read:

Inside the store, Courtney found a jacket on a rack with a sign that said:

The regular price of the jacket was $120. If Courtney buys the jacket during the sale, how much will she pay?

4. Last year, a football player averaged 150 yards rushing per game. This year, his average dropped to 120 yards per game. What was the percent of decrease from last year to this year?

5. In the past, Monica earned $480 per week. Recently, her boss gave her a 7% raise. How much will she earn per week now?

6. Two years ago, the town of Spencer had a population of 2,600. Last year, the population increased by 15%. This year, it increased another 10%. What is the population now?

7. Amber bought a new sofa bed for $980. She paid 7% sales tax. What was the total cost of the purchase, including tax?

8. Charles started with a $5,000 asking price for his pickup truck, but he had to come down to $4,200 to sell it. By what percent did he decrease the price of the truck?

Answers and explanations start on page 209.
For more practice with percents, see page 178.

Using Proportions to Solve Problems

As you saw in the video, you might be using proportion to solve problems without realizing it. Any time you use rates, such as miles per gallon or cost per unit to solve a problem, you are using proportion.

Multiple-Choice Questions

To be successful on the GED Math Test, you need to know how to use proportions to solve multiple-choice questions. Remember, incorrect answer choices on the GED are based on common errors such as mistakes in setting up or solving a proportion.

EXAMPLE: When John works 8 hours per day, he earns $84. At the same rate, how much will he earn if he works only 6 hours per day?

(1) $10.50   (incorrect; shows John's hourly rate: $\frac{\$84}{8}$)

(2) $14.00   (incorrect; divides the amount John earned in an 8-hour day by 6)

(3) $48.00   (incorrect; only multiplies the hours worked: 8×6)

(4) $63.00   (correct; solves the proportion $\frac{\$84}{8} = \frac{\$x}{6}$)

(5) $504.00   (incorrect; shows only the first step in solving the proportion above: $\$84 \times 6$)

Answer: Option **(4) $63.00** is the correct answer. Set up a proportion comparing dollars to hours. To solve the proportion $\frac{\$84}{8} = \frac{\$x}{6}$, multiply $\$84 \times 6$ and divide by 8.

Set-up Questions with Proportions

Some GED questions require you to choose an expression that shows one way of setting up the solution to a problem without actually solving it.

EXAMPLE: The population of Carlsbad is growing at the rate of about 200 people every 3 years. Which expression shows how much the population will grow in 15 years if the growth rate remains the same?

(1) 3×15

(2) $\frac{200}{3}$

(3) $\frac{200}{15}$

(4) $\frac{3 \times 15}{200}$

(5) $\frac{200 \times 15}{3}$

Answer: Option (5) $\frac{200 \times 15}{3}$ is the correct set-up. To solve the proportion $\frac{200}{3} = \frac{x}{15}$, you would multiply 200×15 and divide by 3.

USING PROPORTIONS TO SOLVE PROBLEMS

Choose the <u>one best answer</u> to each question below.

1. The scale on a map is 3 inches = 500 miles. What is the actual distance between two towns that are 7 inches apart on the map?

(1) $\frac{3 \times 7}{500}$

(2) $500 \times 7 \times 3$

(3) $\frac{3}{500 \times 7}$

(4) $\frac{500}{3}$

(5) $\frac{500 \times 7}{3}$

2. At the farmers' market, Linda averages $300 over 5 days of selling plants. If Linda continues to sell at that same rate, how much will she earn in 30 days?

(1) $60

(2) $150

(3) $1,500

(4) $1,800

(5) $9,000

3. The property tax rate in one city is $0.28 per $1,000. If a house is valued at $92,000, which expression can be used to find the property tax on the house?

(1) $\frac{\$0.28 \times \$1,000}{\$92,000}$

(2) $\frac{\$92,000}{\$1,000 \times \$0.28}$

(3) $\frac{\$0.28 \times \$92,000}{\$1,000}$

(4) $\frac{\$1,000 + \$92,000}{\$0.28}$

(5) $\frac{\$0.28 \times \$1,000}{\$1,000 + \$92,000}$

4. A recipe calls for $1\frac{1}{2}$ cups of flour to make 3 dozen cookies. Which expression could be used to find how many cups of flour are needed to make 2 dozen cookies?
(*Hint:* Use 1.5 for $1\frac{1}{2}$.)

(1) $\frac{1.5 \times 3}{2}$

(2) $\frac{1.5 \times 2}{3}$

(3) $\frac{3 \times 2}{1.5}$

(4) $\frac{2 + 3}{1.5}$

(5) $\frac{1.5 \times 2}{1.5 \times 3}$

5. On a long road trip, Marcy drove on stretches of highway with a posted speed limit of 65 miles per hour. If she drove for 3 hours at that speed limit, which expression shows how many miles she traveled?

(1) $\frac{65 \times 3}{1}$

(2) $\frac{65}{1 \times 3}$

(3) $65 + 3 \times 1$

(4) $\frac{3}{65}$

(5) $65 - 3 \times 1$

6. If 3 frozen juice mixes cost $7.50, how much would 6 juice mixes cost?

(1) $7.50

(2) $10.50

(3) $15.00

(4) $22.50

(5) $45.00

Answers and explanations start on page 209.

Percents on the Calculator and Grid

When you use a calculator to solve percent problems, you still have to identify the part, the whole, and the rate. To keep things simple, you can still use proportion to solve percent problems on a calculator. Set up the problem as usual, but do the calculations on the calculator.

Calculator Operations

FINDING THE PART: What is 33% of $20?

You know the rate (33%) and the whole ($20).

Step 1. Set up the proportion as usual: $\frac{\$x}{\$20} = \frac{33}{100}$.

Step 2. Enter 20 $\boxed{\times}$ 33 $\boxed{\div}$ 100 $\boxed{=}$.
The display shows $\boxed{\qquad 6.6}$

Answer: Write the display as dollars and cents: **$6.60.**

FINDING THE RATE: 150 is what percent of 300?

You know the part (150) and the whole (300).

Step 1. Set up the proportion: $\frac{150}{300} = \frac{x}{100}$.

Step 2. Enter 150 $\boxed{\times}$ 100 $\boxed{\div}$ 300 $\boxed{=}$. The display shows $\boxed{\qquad 50.}$

Answer: Write the display over 100 and then as a percent: $\frac{50}{100} =$ **50%.**

FINDING THE WHOLE: 5 is 25% of what number?

You know the part (5) and the rate (25%).

Step 1. Set up the proportion: $\frac{5}{x} = \frac{25}{100}$.

Step 2. Enter 5 $\boxed{\times}$ 100 $\boxed{\div}$ 25 $\boxed{=}$. The display shows $\boxed{\qquad 20.}$

Answer: 5 is 25% of **20.**

EXAMPLES: Solve each example using your calculator and the appropriate steps shown above. Compare your answer to the display.

What is 25% of 44? $\boxed{\qquad 11.}$

8 is what percent of 50? $\boxed{\qquad 16.}$ Write the answer as 16%.

12 is 5% of what number? $\boxed{\qquad 240.}$

Grid Basics

For grid problems with percents, you may be asked to find the part or the whole.
Fill in the grid the same way you would for whole numbers, fractions, and decimals.

EXAMPLE: Out of 365 orders, Anna has filled 20%.
How many orders does she have left to fill?

Anna still has to fill 100% − 20% = 80% of the orders.
Set up a proportion, and solve using your calculator.

$$\frac{x}{365} = \frac{80}{100}$$ $365 \times 80 \div 100 = \mathbf{292}$

The answer to this example is marked on the grid at right.

GED MATH PRACTICE

PERCENTS ON THE CALCULATOR AND GRID

Use your calculator to solve each problem below.

1. 63 is what percent of 180?
2. What percent of 48 is 3?
3. 27 is 0.2% of what number?
4. What is 35% of 400?
5. 100 is 5% of what number?
6. What is 110% of 25?

7. Clint paid $16.74 in sales tax on a digital camera that costs $279.00. What is the sales tax rate?
   - **(1)** 6%
   - **(2)** 6.5%
   - **(3)** 7%
   - **(4)** 7.25%
   - **(5)** Not enough information is given.

Mark your answers on the grids provided. You can use your calculator.

Questions 8 and 9 refer to the chart below.

POTTED PLANT SALE	
4-inch pot	$1.99
6-inch pot	$3.49
8-inch pot	$4.99

8. At 6.5%, how much tax would there be on one 6-inch pot and two 4-inch pots of herbs? (Round to the nearest cent.)

9. The prices above are 15% off the original price. What was the price of an 8-inch pot before the sale?

Answers and explanations start on page 209.
For more practice with calculators and grids, see page 179.

GED Review: Ratio, Proportion, and Percent

Part I: Choose the <u>one best answer</u> to each question below. You <u>may</u> use your calculator.

1. The owners of Zia Gift Shop estimate that its revenues will be $45,000. How much will the store spend on inventory if it usually spends 25% of revenues?
   (1) $11,250
   (2) $9,000
   (3) $6,750
   (4) $5,250
   (5) $4,500

2. This year, rent for the gift shop is $13,500. After a rent increase next year, the rent will be $14,850. What is the percent of increase in the rent?
   (1) 10%
   (2) 9%
   (3) 8%
   (4) 7%
   (5) Not enough information is given.

3. Last year, a sports group spent $350 of a $5,000 budget on awards. If next year's budget is $7,000, what will it spend on awards if it devotes the same percent of the budget?
   (1) $250
   (2) $275
   (3) $315
   (4) $400
   (5) $490

4. George earned $68 for 8 hours of work. At that rate, how much would he earn for 168 hours of work?
   (1) $21
   (2) $168
   (3) $544
   (4) $1,428
   (5) $11,424

5. Cara wants to make a poster by enlarging a photo that measures 5 inches tall by 7 inches wide. If the height of the poster will be 17.5 inches, how many inches wide will it be?

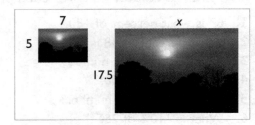

Part II: Choose the <u>one best answer</u> to each question below. You <u>may not</u> use your calculator.

6. Vanessa's last paycheck showed $120 in deductions. Her take-home pay was $480. The deductions are what percent of her TOTAL pay?

   (1) 20%
   (2) 21%
   (3) 23%
   (4) 25%
   (5) 27%

7. William received a $5,000 car loan with a simple interest rate of 8% per year. If he borrows the money for two years, how much interest will he pay? (*Hint:* First find the amount of interest for one year.)

   (1) $200
   (2) $400
   (3) $600
   (4) $800
   (5) $1,000

8. A coat that normally sells for $80 is on sale for $70. What is the percent decrease in price?

   (1) 0.125%
   (2) 1.25%
   (3) 10.0%
   (4) 12.5%
   (5) 125%

9. Hummingbird food can be made using a ratio of 1 cup sugar to 4 cups water. Which expression can be used to find how much sugar is needed to make 1 gallon of hummingbird food? (*Hint:* 1 gallon = 16 cups)

   (1) $\dfrac{1 \times 4}{16}$
   (2) $\dfrac{1 \times 16}{4}$
   (3) $\dfrac{4 \times 16}{1}$
   (4) $\dfrac{4 \times 4}{16}$
   (5) $\dfrac{1 + 16}{4}$

Question 10 refers to the map below.

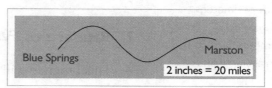

Blue Springs   Marston
2 inches = 20 miles

10. If Blue Springs and Marston are 2.5 inches apart on the map, what is the actual distance between the 2 towns?

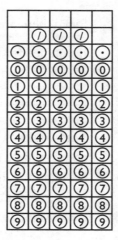

Answers and explanations start on page 210.

Measurement

1. Think About the Topic

The program that you are going to watch is about *Measurement*. The video will show different measurement units, how they're used, and how to convert common units. You will see people using measurement tools and applying what they find. In the course of daily life, you develop your own ways of measuring things you need. Let's see how you can use that knowledge on the GED Math Test.

2. Prepare to Watch the Video

This program will help develop your understanding of measurement tools and measurement units. Answer the questions below to see how much you already know.

How many feet equal 36 inches?

_____

Since there are 12 inches in 1 foot, 36 inches equals 3 feet.

If you needed to measure the weight of a large breed of dog, which unit would you use—ounces, pounds, or tons?

_____

You would measure the weight of a large dog in pounds. The dog would be too heavy to weigh using ounces but not heavy enough to weigh in tons. Having a sense of the size of different measurement units will help you at home, at work, and on the GED Math Test.

LESSON GOALS

MATH SKILLS

- Use standard and metric measurements
- Work with perimeter, area, and volume
- Find perimeter and area of irregular shapes

GED PROBLEM SOLVING

- Draw a picture

GED MATH CONNECTION

- GED formulas page

GED REVIEW

EXTRA PRACTICE PP. 181–183

- Find Perimeter, Area, and Volume
- Find Perimeter and Area of Irregular Shapes
- Use the GED Formulas Page

3. Preview the Questions

Read the questions under *Think About the Program* below, and keep them in mind as you watch the program. You will be reviewing them after you watch.

4. Study the Vocabulary

Review the terms to the right. Understanding the meaning of math vocabulary will help you understand the video and the rest of this lesson.

WATCH THE PROGRAM

As you watch the program, pay special attention to the host who introduces or summarizes major ideas that you need to learn about. The host will also provide important information about the GED Math Test.

AFTER YOU WATCH  25

1. Think About the Program

What types of measuring tools are used in cooking?

How do you find the perimeter of a figure (the distance around the outside of a figure)?

Which system makes it easier to convert units—metric or standard? Why?

2. Make the Connection

The program talked about using measuring tools. Think about the tools you use to measure length, time, weight, and capacity. Which measuring tools do you use most often?

area—the measure of the surface of a figure; measured in square units

equivalency—a statement that two amounts are equal; for example, 1 ft. = 12 in.

gram—the basic metric unit of weight

irregular figure—a figure made by combining several shapes

liter—the basic metric unit of volume, or capacity

meter—the basic metric unit of length

perimeter—a measure of the distance around a shape or object

powers of ten—a number with a 1 followed by any number of zeros; 10, 100, 1,000, and so on

rectangle—a four-sided figure with four right angles and opposite sides of equal length

rectangular container—a box shape

right angle—a square corner like the corner on a sheet of paper

square—a shape with four sides of equal measure and four right angles

volume—the measure of the space inside a three-dimensional object; measured in cubic units

"Before measuring anything, look at your measuring tool. Make sure you understand the markings."

Measurement Units

The English System of Measurement

The measurement system used in the United States is called the English system. The chart below shows the common units of measure in the English system and their equivalencies. The **equivalencies** show how the units are related.

Notice that each unit has been given a benchmark. A **benchmark** is a common object that you can picture to remember the size of a unit.

Learning benchmarks can develop your common sense about measurement. If the benchmark in the table doesn't work for you, make up one of your own. The best benchmarks come from your own experience.

	Unit	Equivalencies	Benchmark
Length	inch (in.) foot (ft.) yard (yd.) mile (mi.)	1 ft. = 12 in. 1 yd. = 3 ft. = 36 in. 1 mi. = 5,280 ft.	the width of two fingers the length of a shoe the length of your stride eight city blocks
Weight	ounce (oz.) pound (lb.) ton	1 lb. = 16 oz. 1 ton = 2,000 lb.	the weight of a utility bill the weight of a can of soup the weight of a small car
Volume	fluid ounce (fl. oz.) cup (c.) pint (pt.) quart (qt.) gallon (gal.)	1 c. = 8 fl. oz. 1 pt. = 2 c. 1 qt. = 2 pt. 1 gal. = 4 qt.	two tablespoons a small coffee cup a small carton of ice cream a large jar of mayonnaise a large plastic jug of milk

The units of time and their abbreviations are used the world over. You will need to memorize these equivalencies and those in the table above to do well on the GED Math Test.

Time Equivalencies
1 minute (min.) = 60 seconds (sec.)
1 hour (hr.) = 60 min.
1 day = 24 hr.
1 week = 7 days
1 year (yr.) = 12 months (mo.) = 365 days

To change from one unit to another, you can use the proportion skills that you learned in Program 24. The first ratio is the equivalency that shows the relationship between the two units. Make sure the second ratio is written with the labels in the same order.

Use x in the proportion to represent the value that you are trying to find.

E X A M P L E : A board is 6 feet long. How many <u>inches</u> long is the board?

Step 1. Write a proportion, using the fact 1 ft. = 12 in. $\frac{1\text{ ft.}}{12\text{ in.}} = \frac{6\text{ ft.}}{x\text{ in.}}$

Step 2. Solve: $6 \times 12 = 72$; $72 \div 1 = 72$ inches

Answer: 6 feet = **72 inches**

If you understand the relationship between feet and inches, you can probably see that multiplying 6 feet by 12 also equals 72. As a general rule, <u>multiply</u> to change from a <u>larger unit (feet) to a smaller one (inches)</u>, and <u>divide</u> to change from <u>smaller to larger</u>. However, if this idea seems confusing, use proportion. It will always work.

E X A M P L E : Gary is making lemonade from a powdered mix. The can's label says that the entire container will make 14 quarts. How many <u>gallons</u> can he make from the can?

Step 1. Write a proportion, using the fact 1 gal. = 4 qt. $\frac{1\text{ gal.}}{4\text{ qt.}} = \frac{x\text{ gal.}}{14\text{ qt.}}$

Step 2. Solve: $1 \times 14 = 14$; $14 \div 4 = 3.5$ or $3\frac{1}{2}$ gallons

Answer: Gary can make **3.5**, or $3\frac{1}{2}$, **gallons** of lemonade.

MEASUREMENT · PRACTICE I

A. Convert the measurements as directed.

1. How many weeks are in 105 days?
2. How many cups are equal to 12 pints?
3. How many inches equal $3\frac{3}{4}$ feet?
4. How many cups are in 32 fluid ounces?
5. How many inches are in 3 yards?
6. How many feet are in 2 miles?
7. How many years are in 30 months?
8. How many ounces are in $4\frac{1}{2}$ pounds?
9. How many fluid ounces are in a pint? (*Hint:* You need two steps. Change to cups and then pints.)

B. Use benchmarks to choose the best answer.

10. Which unit would most likely be used to measure the weight of a cell phone?
    (1) inches
    (2) ounces
    (3) pounds

11. Which unit would most likely be used to measure the amount of milk needed to make pancakes?
    (1) pounds
    (2) cups
    (3) quarts

12. Which unit would most likely be used to measure how much a barrel will hold?
    (1) fluid ounces
    (2) gallons
    (3) tons

13. Which unit would most likely be used to measure the length and width of a photograph?
    (1) inches
    (2) ounces
    (3) feet

Answers and explanations start on page 210.
For more practice with measurement, see page 180.

The Metric System

The **metric system** is a scientific system of measurement based on our decimal system of numbers. It is used throughout the world because it is easier to use and understand than the English system of measurement, since it is based on multiples of 10.

The metric system has three basic units of measure. The **meter** is used to measure length, the **gram** is used to measure weight, and the **liter** is used to measure volume. Study the benchmarks below to gain an understanding of the size of these units.

The Basic Metric Units	
meter (m)	a little longer than a yardstick, a very long stride
gram (g)	much smaller than an ounce, the weight of a small paper clip
liter (L)	half of a 2-liter bottle of soda pop, a little more than a quart

You can use these basic units to make larger and smaller units by multiplying and dividing by 10, 100, and 1,000. You can form the names for other units by adding prefixes to the beginning of the basic units.

kilo- means 1,000   *centi-* means $\frac{1}{100}$ *milli-* means $\frac{1}{1,000}$

Study these examples to see how the prefixes are used:

- A kilometer is 1,000 meters, a little more than half a mile.
- A centimeter is $\frac{1}{100}$ of a meter, a little less than half an inch.
- A millimeter is $\frac{1}{1,000}$ of a meter, about the width of a pencil point.

The prefixes will help you understand the size of metric units. You can also memorize the basic equivalencies below.

Length	1 meter (m) = 100 centimeters (cm) = 1,000 millimeters (mm)
	1 centimeter (cm) = 10 millimeters (mm)
	1 kilometer (km) = 1,000 meters (m)
Weight	1 gram (g) = 1,000 milligrams (mg)
	1 kilogram (kg) = 1,000 grams (g)
Volume	1 liter (L) = 1,000 milliliters (mL)
	1 kiloliter (kL) = 1,000 liters (l)

You can use proportion to make conversions with metric measurements.

E X A M P L E : A pipe is 350 centimeters long. How many <u>meters</u> in length is the pipe?

Step 1. Write a proportion using the fact 1 m = 100 cm. $\frac{1\,m}{100\,cm} = \frac{x\,m}{350\,cm}$

Step 2. Solve: $1 \times 350 = 350$; $350 \div 100 = 3.5$ meters

Answer: The pipe is **3.5 meters** long.

You can also change from one metric unit to another by multiplying or dividing by a **power of ten** such as 10, 100, or 1,000.

Use these shortcuts to multiply or divide quickly by a power of ten.

- Count the number of zeros in the power of ten. This is the number of places you will move the decimal point.
- To multiply, move the decimal point to the *right*. To divide, move it to the *left*.

E X A M P L E : A large jar of peanut butter weighs 1.08 kilograms. How many grams does the jar of peanut butter weigh?

Step 1. Write a proportion using the fact 1 kg = 1,000 g. $\frac{1\,kg}{1,000\,g} = \frac{1.08\,kg}{x\,g}$

Step 2. Solve: $1.08 \times 1,000 = 1,080 = 1,080$
$1,080 \div 1 = 1,080$ grams

Answer: The jar of peanut butters weighs **1,080 grams.**

MEASUREMENT ▪ PRACTICE 2

A. Convert the measurements as directed.

1. How many kilograms are in 1,500 grams?

2. How many meters are equal to 2.4 kilometers?

3. How many millimeters are in 3.7 meters?

4. How many kilometers are in 3,250 meters?

5. How many centimeters are in 4 meters?

6. How many kiloliters are in 4,000 liters?

7. How many millimeters are in 18 centimeters?

8. How many grams are equal to 12.25 kilograms?

B. Use benchmarks to choose the best answer.

9. Which unit would most likely be used to express the weight of a dime?
   (1) millimeters
   (2) grams
   (3) kilograms

10. Which unit would most likely be used to express the height of a bookcase?
    (1) millimeters
    (2) kilometers
    (3) meters

Answers and explanations start on page 210.
For more practice with metric units, see page 180.

Measurement Operations

Addition

English   To add English units of measure, we line up like units and add. You may need to regroup to simplify your answer.

E X A M P L E : Kristen has two pieces of molding. One is 4 feet 10 inches long. The other measures 3 feet 8 inches. What is the total length of the molding?

Step 1. Line up the columns and add:

$$\begin{array}{r} 4 \text{ ft. } 10 \text{ in.} \\ + \ 3 \text{ ft. } \ 8 \text{ in.} \\ \hline 7 \text{ ft. } 18 \text{ in.} \end{array}$$

Step 2. Use the fact 1 ft. = 12 in. to change 18 in. to feet and inches: 18 in. ÷ 12 = 1 ft. 6 in. Simplify the answer: 7 ft. 18 in. = 7 ft. + 1 ft. 6 in. = 8 ft. 6 in.

Answer: Kristen has **8 feet 6 inches** of molding.

Metric   In the metric system, operations are much easier to perform. Make sure all the measurements you need to add are written in the same unit. Then line up the decimal points and add.

E X A M P L E : Marc has two bottles of a photo-processing chemical. One bottle holds 1.6 liters. The other has 0.85 liters. How many liters of the chemical does Marc have in all?

Line up the decimal points and add:

$$\begin{array}{r} 1.6 \\ + \ 0.85 \\ \hline 2.45 \end{array}$$

Answer: Marc has **2.45 liters** of the chemical.

Subtraction

English   You may need to regroup to subtract units using the English system of measurement.

E X A M P L E : A shipping box can hold 5 pounds. An object weighing 2 pounds 10 ounces is put in the box. How much more weight can be added to the box?

Line up the columns. Since the top number has no ounces, regroup. Borrow 1 from the pound column and change it to ounces. Use the fact 1 pound = 16 ounces. Then subtract.

$$\begin{array}{r} 5 \text{ lb.} \\ - \ 2 \text{ lb. } 10 \text{ oz.} \end{array} \quad = \quad \begin{array}{r} 4 \text{ lb. } 16 \text{ oz.} \\ - \ 2 \text{ lb. } 10 \text{ oz.} \\ \hline 2 \text{ lb. } \ 6 \text{ oz.} \end{array}$$

Answer: The box can hold **2 pounds 6 ounces** more weight.

Metric   To subtract metric units, make sure the units are the same, line up the decimal points, and subtract.

E X A M P L E :   A package weighing 10 kilograms is how many kilograms heavier than a package weighing 3,800 grams?

Step 1.   Change grams to kilograms. Write a proportion, $\frac{1\,kg}{1,000\,g} = \frac{x\,kg}{3,800\,g}$, and solve:
$1 \times 3,800 \div 1,000 = 3.8$, so 3,800 grams = 3.8 kilograms.

Step 2.   Line up the decimal points and subtract:
$$\begin{array}{r} 10.0 \\ -\ \ 3.8 \\ \hline 6.2 \end{array}$$

Answer: The 10-kilogram package weighs **6.2 kilograms** more.

MEASUREMENT ▪ PRACTICE 3

A.   Solve as directed.

1.   Add 4 ft. 6 in. and 6 ft. 10 in.
2.   Subtract 40 min. from 1 hr. 30 min.
3.   Add 3.5 L and 0.9 L.
4.   Subtract 1.54 m from 7 m.
5.   Find the sum of 3 lb. 14 oz. and 11 lb. 15 oz.

6.   How much longer is 2 yd. than 1 ft. 4 in.?
7.   What is the total of 5.6 kg and 1,550 g?
8.   How many centimeters less is 48 cm than 1 m?

B.   Solve.

9.   Cooper spent the following amounts of time studying for a test: 1 hr. 45 min., 20 min., 1 hr. 30 min., and 45 min. What was the total time that he spent studying?

10.   On Monday, Nita used three pints of liquid adhesive from a 1-gallon can. How many <u>pints</u> of adhesive are left in the can?

11.   Zane bought a 12-yd. roll of screen material to replace two window screens. He uses 4 ft. 6 in. for the first screen and 3 ft. 9 in. for the second screen. How much is left on the roll? Express your answer in feet and inches.

12.   A weight of 3,500 grams is how many more grams than 2.8 kilograms?

13.   Mira ordered three lamps from Canada. The shipping weights of the lamps were 3.64 kg, 5.4 kg, and 0.375 kg. What was the total weight of the shipment?

14.   Carrie has a metal rod that is 1.8 meters long. If she cuts off the shaded piece shown below, how many <u>meters</u> will be left?

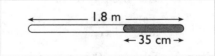

15.   James can get free shipping if his order weighs up to 20 pounds. He orders a metal toolbox that weighs 11 lb. 9 oz. If he orders one more item, what is the most it can weigh for James to get free shipping?

Answers and explanations start on page 210.
For more practice converting measurement units, see page 180.

Multiplication

English   A measurement in the English system may be a combination of units. Multiply each part of the measurement separately, and then simplify the answer.

E X A M P L E : The plans for a small bookcase call for six boards, each 2 feet 9 inches long. What is the total length of the boards in feet and inches?

Step 1.   Multiply each unit part:

$$
\begin{array}{r}
2 \text{ ft. } 9 \text{ in.} \\
\times \qquad 6 \\
\hline
12 \text{ ft. } 54 \text{ in.}
\end{array}
$$

Step 2.   Simplify. Convert 54 inches to feet and inches.
12 ft. 54 in. = 12 ft. + 4 ft. 6 in. = 16 ft. 6 in.

Answer: The total length of the boards is **16 ft. 6 in.**

Metric   In the metric system, multiply as you would any decimal numbers.

E X A M P L E : A piece of copper tubing is 2.5 meters. If eight pieces of the tubing are joined, what will be the total length in meters?

Multiply:
$$
\begin{array}{r}
2.5 \\
\times \quad 8 \\
\hline
20.0
\end{array}
$$

Answer: The combined length will be **20 meters.**

Division

English   To divide a measurement in the English system, convert the measurement to the smallest unit used in the problem. Then divide. Simplify your answer if necessary.

E X A M P L E : Donna has 12 lb. 8 oz. of flour. If she divides the flour evenly into five containers, how much flour will she put in each container?

Step 1.   Change 12 lb. 8 oz. to ounces. Find the number of ounces in 12 pounds.
Use the fact 1 lb. = 16 oz. Then add 8 oz.

$$\frac{1 \text{ lb.}}{16 \text{ oz.}} = \frac{12 \text{ lb.}}{x \text{ oz.}} \qquad 16 \times 12 = 192 \qquad 192 \div 1 = 192 \qquad 192 + 8 = 200 \text{ oz.}$$

Step 2.   Divide: 200 oz. ÷ 5 = 40 oz.
Step 3.   Simplify: 40 oz. ÷ 16 = 2 lb. 8 oz.

Answer: Donna will put **2 lb. 8 oz.** in each container.

Metric   Use the rules for dividing decimals to divide a metric measurement.

E X A M P L E : Steve runs a jogging path at the park each day. He knows that four laps of the path are equal to a distance of 2.8 kilometers. How many kilometers is one lap?

Divide:
$$
\begin{array}{r}
.7 \\
4\overline{)2.8} \\
\underline{2\,8} \\
0
\end{array}
$$

Answer: One lap of the jogging path is **0.7 kilometer** long.

A. Solve as directed.

1. What is 1 lb. 4 oz. multiplied by 4?
2. Divide 1 yd. 1 ft. 6 in. into 9 equal lengths. Express your answer in inches.
3. What is 7.5 kg multiplied by 6?
4. What is 12.6 meters divided into 5 equal lengths?

5. Divide 3 gallons 2 quarts by 2.
6. What is the weight in pounds of 8 packages if each weighs 12 ounces?
7. What is 1.8 L divided by 3?
8. What is the length in meters of 24 centimeters multiplied by 10?

B. Solve.

9. Eva is making candles. She needs 10 ounces of hot wax to fill one mold. How many <u>pounds</u> of wax should she buy to make 8 candles?

10. Sean worked 1 hr. 20 min. of overtime every day for 8 days. How much overtime did he work in all during this period?

11. Alice refinishes wood furniture. She buys wood stripper in a 7.5-liter can and then pours it into smaller containers. If she fills 5 containers, how many liters does she put in each container?

12. A book weighs 1.3 kilograms. How much would 6 of the same book weigh?

13. A project in a craft book calls for the following lengths of leather lacing.

Craft Project #4A	
Quantity	**Length**
4	2 ft. 3 in.
8	9 in.

How many <u>feet</u> of leather lacing are needed to complete the project?

14. Patrice bought one gallon of fruit punch for her child's party. She estimates that each paper cup will hold 8 ounces. How many paper cups can she fill with fruit punch?

Answers and explanations start on page 211.
For more practice with measurements, see page 180.

"I adapted a design from a magazine to fit the area and dimensions of my yard."

Perimeter, Area, and Volume

Perimeter

The **perimeter** of a shape is the distance around the shape. To find the perimeter, you add the measures of all the sides.

E X A M P L E : Nita and Craig plan to fence their yard. The diagram to the right shows the dimensions of their yard. How many feet of fencing do they need?
Add the four sides: 112 + 170 + 105 + 210 = 597 feet
Answer: Nita and Craig will need **597 feet** of fencing.

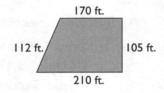

Some shapes have special properties that make it easy to find the perimeter. A **square** has 4 sides of equal length and 4 right angles. A **right angle** is like the corner of a piece of paper. You can multiply the measure of one side by 4 to find the perimeter.

E X A M P L E : What is the perimeter of a square if one side measures 8 inches?
Multiply by 4: 8 in. × 4 = 32 in. Simplify: 32 in. ÷ 12 = 2 ft. 8 in.
Answer: The perimeter of the square is **2 ft. 8 in.**

A **rectangle** is a four-sided shape with right angles that has opposite sides of the same length. A rectangle has two dimensions: length and width. To find the perimeter of a rectangle, multiply the length by 2 and the width by 2, and add.

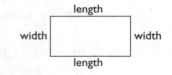

E X A M P L E : What is the distance around the flowerbed?
Multiply the length by 2: 6 ft. × 2 = 12 ft.
Multiply the width by 2: 4 ft. × 2 = 8 ft. Add: 12 ft. + 8 ft. = 20 ft.
Answer: The perimeter of the flowerbed is **20 ft.**

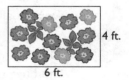

Area

Area is the measure of the surface of a figure. Area could tell you how much paint to buy to cover a wall or how many tiles to buy to cover a floor. To find area, you must figure out how many square units will fit inside the shape. Multiply the length and the width to find the area of a rectangle in square units.

E X A M P L E : What is the area of the flowerbed shown at right?
Multiply the length and width: 6 × 4 = 24 sq. ft.
Answer: The area of the flowerbed is **24 sq. ft.** In other words, 24 squares, each measuring 1 foot per side, could fit inside.

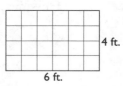

Volume

Volume is the measure of space inside a three-dimensional object. When something has three dimensions, it has length, width, and height or depth. Imagine a box that is 1 foot on each side. We say that the box takes up 1 cubic foot of space. Now imagine filling a closet with these boxes. The number of boxes that will fit in the closet is the volume of the closet. Volume is measured in cubic units.

A closet is an example of a rectangular container. A **rectangular container** has sides called faces that are all rectangles or squares. To find the volume of a rectangular container, multiply the length by width by height (or depth).

E X A M P L E : What is the volume of the box?

Multiply: 12 in. × 10 in. × 5 in. = 600 cubic inches

Answer: The volume of the box is **600 cubic inches.** In other words, the box will hold 600 cubes that measure 1 inch on each edge.

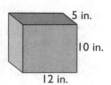

MEASUREMENT ▪ PRACTICE 5

A. Find the perimeter and area of each figure.

1.

2.5 cm

2.5 cm

2.

3 ft.

5 ft.

3.

7.8 cm

2 cm

B. Find the volume of each box.

4.
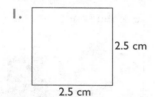
3 cm

7 cm

11 cm

5.
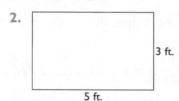
2 ft.

8 ft.

2.5 ft.

6.
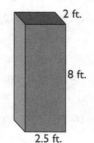
5 in.

5 in.

5 in.

C. Solve.

7. Amy plans to put crown molding around the top of the walls in her dining room. If the rectangular room is 13 feet long and 12 feet wide, how many feet of molding will she need?

8. The floor of a rectangular storage shed is 7 feet long and 4 feet wide. If the shed is 5 feet in height, what is the volume of the shed in cubic feet?

Answers and explanations start on page 211.
For more practice finding perimeter, area, and volume, see page 181.

Irregular Figures

In real life, figures are often a combination of shapes. To find the perimeter of an irregular figure, you may need to find the lengths of unmarked sides.

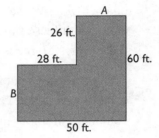

E X A M P L E : A diagram of the lobby of an office building is shown to the right. What is the perimeter of the lobby?

Look at the drawing. Some of the measurements are missing, but you can find these measurements by performing a few simple calculations.

Step 1. Find the measure of the wall marked *A*. You know the sum of both north walls must equal the length of the south wall.
28 ft. + _____ = 50 ft. Subtract: 50 ft. − 28 ft. = 22 ft.

Step 2. Find the measure of the wall marked *B*. You know the sum of both west walls must equal the length of the east wall.
26 ft. + _____ = 60 ft. Subtract: 60 ft. − 26 ft. = 34 ft.

Step 3. Now find the perimeter by adding the measures of all the walls:
60 + 50 + 34 + 28 + 26 + 22 = 220 ft.

Answer: The perimeter of the lobby is **220 feet.**

To find the area or volume of an irregular shape, you need to break the shape into smaller regular shapes.

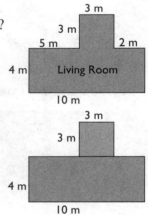

E X A M P L E : What is the area of the living room shown to the right?

Step 1. The living room is formed from two shapes: a square and a rectangle. The second drawing shows how the room can be divided into two shapes and drops out any unnecessary measurements. Now you can easily find the area of each shape.
Area of square: 3 × 3 = 9 square meters
Area of rectangle: 10 × 4 = 40 square meters

Step 2. Add to find the total area: 9 + 40 = 49 square meters

Answer: The area of the living room is **49 square meters.**

To solve some problems, you will need to use facts from the problem and facts from a diagram. Think about what you are trying to find, and then gather the facts you need.

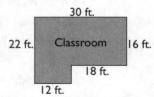

E X A M P L E : A contractor needs to find the volume in cubic feet of a classroom in order to install air conditioning. What is the volume of the classroom shown in the diagram if all walls are 9 feet in height?

Step 1. The classroom can be divided into two rectangular sections. You are solving for volume, so you will need the length, width, and height of both sections. You can get the length and width from the diagram. The problem says that the height is 9 feet.

Step 2. Find the volume of the sections separately.
Section A: $22 \times 12 \times 9 = 2{,}376$ cubic feet
Section B: $18 \times 16 \times 9 = 2{,}592$ cubic feet

Step 3. Add to find total volume:
$2{,}376 + 2{,}592 = 4{,}968$ cubic feet

Answer: The volume of the classroom is **4,968 cubic feet.**

There is often more than one way to divide a complex figure into smaller regular shapes. Choose the way that seems best to you. The answer will be the same no matter how you divide the figure.

MEASUREMENT ▪ PRACTICE 6

Solve.

1. Art plans to put new tile in his kitchen. The diagram below shows the dimensions of the kitchen.

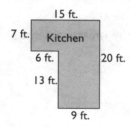

How many square feet of tile will he need?

Questions 2 and 3 refer to the diagram below.

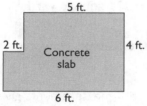

2. A contractor plans to pour a concrete slab in the shape shown below. If the slab is $\frac{1}{2}$ foot thick, how many cubic feet of concrete will the contractor need to mix?

3. How many feet of wood will the contractor need to build a frame for the slab? (*Hint:* Find the perimeter.)

4. In a fitness center, the floor surrounding a small swimming pool needs to be resurfaced. The area to be resurfaced is shown by the gray shaded section below.

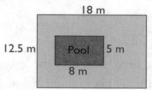

How many square meters of floor need to be resurfaced? (*Hint:* Subtract the area of the pool from the area of the entire floor.)

5. A toy manufacturer plans to use the box below for a new game.

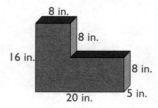

In cubic inches, how much will the box hold?

Answers and explanations start on page 211.
For more practice with measurement, see page 182.

Draw a Picture

As you saw in the video, our system of measuring length, weight, and volume is part of the English system of measurement. When working with measurements, you can draw a picture to make solving problems easier.

Draw a picture to help organize information given in a problem and to identify what the problem is asking you to do.

When taking the GED Math Test, you will be given scratch paper to use to work out the problems. Draw a picture on the scratch paper, and label your drawing with numbers given in the problem.

> **E X A M P L E :** A room measures 12 feet by 16 feet. How many <u>square yards</u> of carpeting are needed to cover the floor?
>
> Draw a rectangle and label the length and width with the measurements given in the problem.

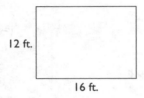

12 ft.

16 ft.

To find the area, multiply the length and width: 12 feet × 16 feet = 192 square feet. Next find the number of square yards in 192 square feet.
Using the fact 1 square yard = 9 square feet, divide: $\frac{192}{9} = 21\frac{1}{3}$ **square yards.**

> **E X A M P L E :** A 10-foot by 10-foot storage unit is 8 feet tall. How many cubic feet of storage space does the unit contain?
>
> Draw a box and label the length, width, and height with measurements given in the problem.

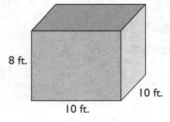

8 ft.

10 ft.

10 ft.

To find the volume, multiply the length by width by height:
10 feet × 10 feet × 8 feet = **800 cubic feet.**

Problems with Irregular Shapes

When solving problems with irregular shapes, you may need to draw a more complex picture and do more than one calculation to solve the problem. Draw and label the picture on scratch paper, and write down each calculation.

EXAMPLE: The Bowens built a 3-foot-wide deck around a rectangular swimming pool. The pool measures 12 feet long by 9 feet wide. What is the area of the deck?

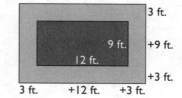

Step 1. Draw and label a picture of the pool and the deck.

Step 2. Find the area of the pool:
12 feet × 9 feet = 108 square feet.

Step 3. Find the area of the deck and pool:
(12 + 6) × (9 + 6) = 270 square feet.

Step 4. Subtract: 270 − 108 = **162 square feet**.

 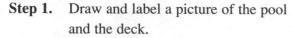

DRAW A PICTURE

Draw a picture to help solve each problem.

1. Shelly is building a railing around her deck. The deck is attached to her house on one of the long sides. How long will the railing be if her deck measures 15 feet by 20 feet?

2. Hector is painting the outside walls of his 8-foot by 8-foot storage shed. Find the surface area to be painted if the shed walls are $7\frac{1}{2}$ feet tall.

3. What is the volume in cubic feet of a garden pool that measures 6 feet by 4 feet and is 15 inches deep?

4. Rhonda's cabin has two triangular windows with sides of 3 feet, 4 feet, and 5 feet. Write an expression showing the amount of molding needed to go around both windows.

Draw a picture and calculate any missing information. Then solve.

5. The city parks department wants to put a fence around a 30-foot square picnic area, leaving two 4-foot openings on opposite sides for an entrance and an exit. How many feet of fencing will the parks department need?

6. Merissa is making a quilt with 8-inch square blocks. She will have a 4-inch solid border around the quilt. How many blocks are needed to make a quilt that measures 48 inches by 72 inches?

Answers and explanations start on page 211.

Formulas Page

The partial formulas table below is similar to the one on the GED Math Test.

AREA of a:

square	Area = side^2
rectangle	Area = length × width
parallelogram	Area = base × height
triangle	Area = $\frac{1}{2}$ × base × height
circle	Area = π × radius^2; π is approximately equal to 3.14.

PERIMETER of a:

square	Perimeter = 4 × side
rectangle	Perimeter = 2 × length + 2 × width
triangle	Perimeter = side$_1$ + side$_2$ + side$_3$
circumference of a circle	Circumference = π × diameter; π is approximately equal to 3.14.

VOLUME of a:

cube	Volume = edge^3
rectangular solid	Volume = length × width × height
cylinder	Volume = π × radius^2 × height; π is approximately equal to 3.14.

SIMPLE INTEREST	interest = principal × rate × time
DISTANCE	distance = rate × time
TOTAL COST	total cost = (number of units) × (price per unit)

Formula Problems

Some problems that require formulas are straightforward. You solve for the unknown area, perimeter, volume, interest, distance, or cost. Some problems require you to solve for other variables in the formula.

> **EXAMPLE:** Imelda receives 9% simple interest on a $5,000 investment. How much interest will she earn in 3 years?

Use the interest formula:	$i = prt$
Substitute the given numbers:	$i = \$5,000 \times 9\% \times 3$
Solve for the interest (remember, $9\% = \frac{9}{100} = .09$):	$i = \$5,000 \times .09 \times 3$
	$i = \mathbf{\$1,350}$

> **EXAMPLE:** Sylvia drove 250 miles in $4\frac{1}{2}$ hours. What was her average speed?

Use the distance formula:	$d = rt$
Substitute the given numbers:	252 miles = r × 4.5 hours
Solve for the rate:	$\frac{252}{4.5} = r$
	56 miles per hour = r

Set-up Problems

Some of the formula problems on the GED Math Test require you to choose the correct way to set up a problem. You are not required to calculate the answer.

> **E X A M P L E :** The area of a rectangular parking lot is 10,000 square meters. If the length of the parking lot is 125 meters, which expression could be used to find the width?
>
> **(1)** $\frac{125}{10,000}$
>
> **(2)** $125 \times 10,000$
>
> **(3)** $\frac{10,000 - 125}{2}$
>
> **(4)** $\frac{10,000}{125}$
>
> **(5)** $10,000 - 2(125)$

The correct answer is **(4)** $\frac{10,000}{125}$. Use the formula for area of a rectangle, $A = lw$
Substitute: $10,000 = 125 \times w$. Solve for w: $w = \frac{10,000}{125}$.

GED MATH PRACTICE

FORMULAS PAGE

Use a formula to solve each problem below.

1. To compare prices, Leona wants to find the cost per roll in a package of paper towels. What is the cost per roll if she pays $7.59 for a package of 6? Round to the nearest cent.

2. A triangular wall hanging has a base of 3 feet and a height of 3.5 feet. What is the area of the wall hanging?

3. The freezer opening of a refrigerator measures 24 inches by 15 inches. If the capacity of the freezer is 3.75 cubic feet, what is the depth?

4. How long must a piece of ribbon be to just fit around the edge of a piece of needlework in the shape of a circle with a 5-inch diameter?

Choose the expression that could be used to solve the problem.

5. Gabriel is paying off a simple-interest loan of $2,500 at 7% annual interest for 2 years. Which expression could be used to find the TOTAL AMOUNT Gabriel must pay back?
   (1) $2,500 × .07 × 2
   (2) $2,500 + .07 × 2
   (3) $2,500 + ($2,500 × .07 × 2)
   (4) $2,500 + ($2,500 × .07 ÷ 2)
   (5) $2,500 × 7 × 2

6. A regular piece of notebook paper measures 8.5 inches by 11 inches. Which expression could be used to find the area of the paper in SQUARE FEET?
   (1) $\frac{8.5 \times 11}{144}$
   (2) $2(8.5) + 2(11)$
   (3) $\frac{1}{2} \times 8.5 \times 11$
   (4) 8.5×11
   (5) $\frac{8.5 \times 11}{12}$

Answers and explanations start on page 211.
For more practice with formulas, see page 183.

GED Review: Measurement

Part I: Choose the <u>one best answer</u> to each question below. You may refer to the formulas on page 136. You <u>may</u> use your calculator.

1. Terry earned $85 in interest on his investment for one year. If he invested $2,000, what rate of interest did he earn?
   (1) 2.35%
   (2) 2.85%
   (3) 3.5%
   (4) 4.25%
   (5) 5.0%

2. A rectangular shipping crate holds 6 cubic feet. If the length is 1.5 feet and the width is 2 feet, how many feet is the height of the crate?
   (1) 1
   (2) 2
   (3) 3
   (4) 4
   (5) Not enough information is given.

3. A water storage tank is shaped like a cylinder. It has an inside diameter of 100 cm and is 40 cm deep. What is the volume of the tank in cubic centimeters?
   (1)   314,000
   (2)   556,000
   (3)   782,000
   (4) 1,000,000
   (5) 1,256,000

4. A circular driveway encloses a lawn area with diameter of 52 feet. What is the area of the lawn rounded to the nearest square foot?

52 ft.

   (1)   82
   (2)   163
   (3)   2,123
   (4)   6,665
   (5)   8,491

5. A parallelogram with an area of 144 square inches has a base of 12 inches. What is the height in inches?

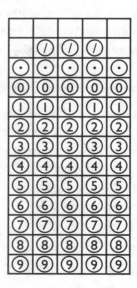

Part II: Choose the <u>one best answer</u> to each question below. You <u>may not</u> use your calculator.

6. Alfredo drove 200 miles in 3 hours. Which expression could be used to find his average speed?

   (1) $\frac{3}{200}$
   (2) 3×200
   (3) $200 + 3$
   (4) $\frac{200}{3}$
   (5) $200 - 3$

7. Dean's horse ran the quarter-mile in 50 seconds. In feet per second, how fast was the horse running?

   (1) 32.8
   (2) 26.4
   (3) 19.3
   (4) 12.5
   (5) 2.0

8. A circle has a radius of 25 cm. How many centimeters is the circumference of the circle?

   (1) 78.5
   (2) 157.0
   (3) 432.5
   (4) 925.0
   (5) 1,962.5

9. How many boxes measuring 12 inches deep by 12 inches wide by 6 inches tall will fit on a store shelf measuring 12 inches deep, 6 feet wide, and 18 inches tall?

   (1) 6
   (2) 9
   (3) 12
   (4) 15
   (5) 18

10. Which of the following expressions could be used to find the cost per can if a 12-pack of soda costs $2.99?

    (1) $\frac{12}{2.99}$
    (2) 12×2.99
    (3) $2.99 + 12$
    (4) $2.99 - 12$
    (5) $\frac{2.99}{12}$

11. What is the perimeter in inches of the figure below?

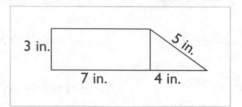

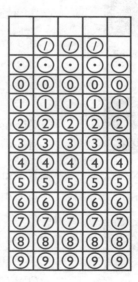

Answers and explanations start on page 212.

Data Analysis

LESSON GOALS

MATH SKILLS

- Use tables
- Read and interpret graphs
- Explore probability

GED PROBLEM SOLVING

- Use the problem-solving method

GED MATH CONNECTION

- Data on the calculator and the grid

GED REVIEW

EXTRA PRACTICE PP. 184–187

- Tables and Graphs
- Probability
- Use the Problem-Solving Method
- Use the Calculator and the Grid with Data

1. Think About the Topic

The program that you are going to watch is about *Data Analysis*. The video will show you different ways that data can be represented. You will see how and when tables and graphs are used. You will also explore finding the mean, median, and mode of a set of data. Knowing what the data means and being able to draw conclusions are important math skills for daily life and the GED Math Test.

2. Prepare to Watch the Video

This program will give ideas about reading and understanding data. Try out your number sense when it comes to data analysis.

If a graph shows a line heading downward, do you think it shows growth or a decline?

_____

On a graph, a line that slopes downward from left to right shows a decline in what is being measured.

If you wanted to find the typical rental cost of one-bedroom apartments in your area, what could you do?

_____

You may have written something like: *I could find the average cost (mean) or the middle cost (median) of rent.* This question is related to finding a value that represents a set of data—a skill you will need for the GED Math Test.

3. Preview the Questions

Read the questions under *Think About the Program* below, and keep them in mind as you watch the program. You will be reviewing them after you watch.

4. Study the Vocabulary

Review the terms to the right. Understanding the meaning of math vocabulary will help you understand the video and the rest of this lesson.

WATCH THE PROGRAM

As you watch the program, pay special attention to the host who introduces or summarizes major ideas that you need to learn about. The host will also tell you important information about the GED Math Test.

AFTER YOU WATCH

26

1. Think About the Program

Which type of graph would be best for comparing values?
(circle, bar, or line graph)

Which type of graph shows parts of a whole?
(circle, bar, or line graph)

Which type of graph shows changes over time?
(circle, bar, or line graph)

2. Make the Connection

The program talked about planning a budget. Which typical value would be best for finding your monthly expenses: mean (average), median (middle number), or mode (most common number)? Explain your reasoning.

---

average—the center of a set of data; see also *mean*

bar graph—a graph that uses bars to compare data

circle graph—a graph that uses sections of a circle to represent the parts of a whole

experimental probability—the ratio of favorable outcomes to the number of times the experiment is performed

line graph—a graph that uses points and lines to show changes in data over time

mean—the sum of a set of numbers divided by the number of values in the set; also called the arithmetic *average*

median—the middle value in a set of data arranged in order from least to greatest or greatest to least

mode—the value that occurs most often in a set of data

pictograph—a graph that uses symbols to represent data

probability—the study of chance, expressed as the ratio of favorable outcomes to possible outcomes

table—an arrangement of data in columns and rows

trend—a pattern of change in data

trial—one performance of an experiment

"As a columnist, I'm always using numbers to explain things to people."

Tables, Charts, and Graphs

Tables and Charts

Information is easier to understand and compare when it is organized. A **table** uses columns and rows to organize information. Tables can be simple, like those used earlier in this book, or more complex, like those found in magazines and at work.

To find information in a table, first read the title and labels carefully. Then find the row and the column where the information you need is located. Finally, find the place where the row and column intersect, or meet.

E X A M P L E : Matt owns shares in three stocks. He checks the price per share for each stock weekly. He made the table below to track prices for the last five weeks.

Matt's Stock Portfolio—Price per Share					
Stocks	June 1	June 8	June 15	June 22	June 29
AOR	$18.09	$16.05	$16.60	$15.50	$13.51
INP	$26.62	$28.18	$21.28	$25.36	$26.75
XON	$34.80	$34.91	$35.47	$36.41	$35.22

If Matt owns 50 shares of INP stock, what was the stock worth on June 22?

Step 1. Find INP in the list of stocks. Follow the INP row across until you are in the column labeled June 22. The box where the row and column meet contains the fact you need. The price per share is $25.36.

Matt's Stock Portfolio—Price per Share					
Stocks	June 1	June 8	June 15	June 22	June 29
AOR	$18.09	$16.05	$16.60	$15.50	$13.51
INP	$26.62	$28.18	$21.28	$25.36	$26.75
XON	$34.80	$34.91	$35.47	$36.41	$35.22

Step 2. Perform the calculations. Matt has 50 shares of INP stock at a price of $25.36 per share. Multiply: $25.36 × 50 = $1,268.00.

Answer: Matt's shares of INP stock were worth **$1,268** on June 22.

Charts use pictures to visually show information and make simple comparisons. There are many different kinds of charts. For example, in a **pictograph**, picture symbols represent numbers. A key tells the value of the symbol.

E X A M P L E : Based on the pictograph, how many workers travel between 10.1 and 15 miles to get to work?

Step 1. Find the facts. There are four "10.1 to 15."

Step 2. Read the key. The key states that each symbol represents 8 workers.

Step 3. Multiply: $4 \times 8 = 32$ workers.

Answer: **32 workers** travel between 10.1 and 15 miles.

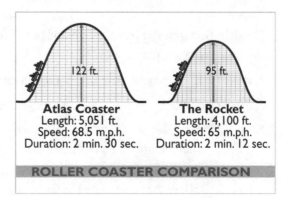

Miles	Number of Workers
Fewer than 5	🚗🚗🚗
5.1 to 10	🚗🚗🚗🚗🚗🚗
10.1 to 15	🚗🚗🚗🚗
More than 15	🚗🚗

🚗 = 8 workers

AVERAGE DISTANCE TRAVELED BY WORKERS FROM HOME TO WORK

DATA ANALYSIS ▪ PRACTICE 1

Solve as directed.

Questions 1 through 3 refer to the table on page 142 titled Matt's Stock Portfolio.

1. Which stock was worth more on June 29 than it was worth on June 8?

2. On June 1, the value of a single share of INP stock was how much greater than a share of AOR stock?

3. Matt sold 80 shares of XON stock on June 29. How much was he paid for his shares?

Questions 4 and 5 refer to the following table.

Life Insurance Monthly Rates				
	$25,000 BENEFIT		**$50,000 BENEFIT**	
AGE	**MALE**	**FEMALE**	**MALE**	**FEMALE**
18–24	$6.90	$5.78	$10.80	$8.56
25–29	$6.90	$5.78	$10.80	$8.56
30–34	$6.90	$5.78	$10.80	$8.56
35–39	$7.34	$6.08	$11.67	$9.17
40–44	$8.70	$7.08	$14.40	$11.15

4. Mary is 27 years old. How much would she pay per month if she buys a $25,000 life insurance policy?

5. A husband and wife are both 38 years old. If each buys a $50,000 policy, how much more will he pay per month than she will?

Questions 6 and 7 refer to the pictograph at the top of this page.

6. How many workers travel 10 or fewer miles? (*Hint:* You will need to add the symbols from two rows.)

7. How many more workers travel 10.1 to 15 miles than travel more than 15 miles to work?

Questions 8 and 9 refer to the chart below.

122 ft.

95 ft.

Atlas Coaster
Length: 5,051 ft.
Speed: 68.5 m.p.h.
Duration: 2 min. 30 sec.

The Rocket
Length: 4,100 ft.
Speed: 65 m.p.h.
Duration: 2 min. 12 sec.

ROLLER COASTER COMPARISON

8. How many feet higher is the first hill of the Atlas than the first hill of the Rocket?

9. The duration is the time it takes to finish the ride. If the rides begin at exactly the same time, which coaster will finish first?

Answers and explanations start on page 212.
For more practice with tables, see page 184.

Using Bar Graphs

A **bar graph** uses bars next to a number scale to show data. A bar graph has two scales, vertical and horizontal. Both the scales are labeled. Read the title of the graph and both sets of labels to understand what the bars represent.

Study the parts of this bar graph.

To find the value that a bar represents, compare the height of the bar with the number scale.

The labels along the bottom show that each bar represents a day of the week.

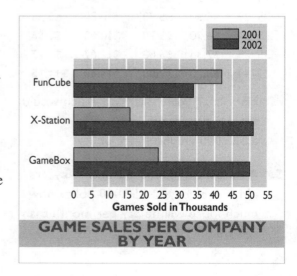

FUTURE BANK AND TRUST WEEK OF FEB. 15

EXAMPLE: A bank manager counts the number of customers each day who come inside the bank for business. She makes the graph shown above. On the day with the fewest customers, <u>about</u> how many customers come into the bank?

Step 1. Examine the bars. The shortest bar is labeled Thursday.

Step 2. Estimate the value of the bar for Thursday. Read across from the top of the bar to the scale on the left. The top of the bar is below 200 customers. The number 190 would be a reasonable estimate.

Answer: About **190 customers** came into the bank on Thursday.

Double-bar graphs have groups of bars. You must read the key to understand the meaning of each bar.

EXAMPLE: About how many more GameBox games were sold in 2002 than FunCube games?

Step 1. Find the facts. The key tells you that the gray bars represent 2002. The GameBox bar for 2002 is at 50 on the scale. The FunCube bar is a little less than 35, or 34.

Step 2. Subtract: $50 - 34 = 16$

Step 3. Interpret the results. The label at the bottom of the graph says that the numbers are in thousands. The number 16 means 16,000.

Answer: GameBox sold about **16,000 more games** than FunCube in 2002.

GAME SALES PER COMPANY BY YEAR

Solve as directed.

Questions 1 through 3 refer to the Future Bank and Trust graph on page 144.

1. About how many customers came into the bank on Wednesday?

2. What was the approximate total of in-bank customers on Friday and Saturday?

3. About how many more customers came to the bank on Monday than on Tuesday?

Questions 4 through 7 refer to the graph below.

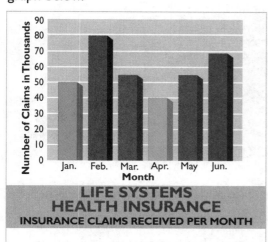

LIFE SYSTEMS HEALTH INSURANCE
INSURANCE CLAIMS RECEIVED PER MONTH

4. About how many claims were received by the company in June?

5. In which two months did the company receive about the same number of claims?

6. How many more claims did the company receive in February than in April?

7. By what percent did the claims increase from January to February?

Questions 8 through 10 refer to the Game Sales graph on page 144.

8. Which company showed a decrease in sales from 2001 to 2002?

9. The three companies sold approximately how many games in 2001?

10. Which company was closest to a 2:1 ratio of 2002 sales to 2001 sales?

Questions 11 through 15 refer to the following graph.

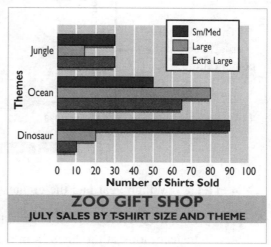

ZOO GIFT SHOP
JULY SALES BY T-SHIRT SIZE AND THEME

11. What theme had the greatest number of extra-large shirts sold?

12. Of the shirts sold, what was the ratio of large dinosaur shirts to large ocean shirts?

13. What percent of the total number of dinosaur shirts sold were small/medium?

14. How many small/medium shirts were sold in all?

15. For which theme did the gift shop sell the fewest shirts?

Answers and explanations start on page 212.
For more practice with tables and graphs, see page 185.

"For most people, the picture conveys more information than the numbers."

Line and Circle Graphs

Using Line Graphs

A **line graph** uses points and lines to show changes in values over time. A line graph has two scales. One scale tells the values of the points. The other shows time intervals such as hours, days, weeks, months, and years.

You can read the points on a line graph using the scales and key as you would the bars on a bar graph. Study the parts of this line graph.

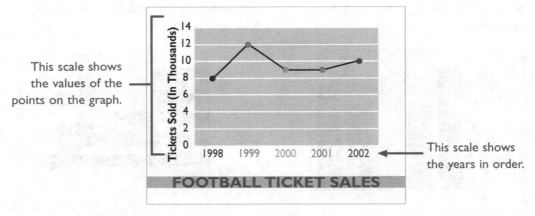

This scale shows the values of the points on the graph.

This scale shows the years in order.

FOOTBALL TICKET SALES

EXAMPLE: Which year had the greatest increase in ticket sales from the year before?

Step 1. Find the increases. Don't spend time finding the values of all of the points. Instead, look at the lines from point to point. If a line goes up, it shows an increase. There are two increases shown on the graph.

Step 2. Compare. Now think of each line as a hill to climb. The line with the steeper slope, the one that would be harder to climb, is the greater increase.

Answer: The greatest increase in ticket sales was in **1999.**

Line graphs can have more than one line. A key will explain the meaning of each line. You can sometimes use line graphs to see **trends,** or patterns, in data.

EXAMPLE: A restaurant's sales of which type of pizza steadily increased from June to September?

Examine both lines. The line for Chicago-style pizza showed small increases and decreases during the time period. The line for stuffed-crust pizza moved upward each month.

Answer: Sales of **stuffed-crust pizza** increased steadily.

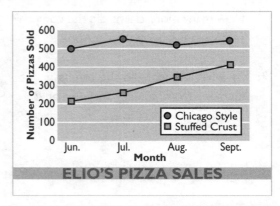

ELIO'S PIZZA SALES

Solve as directed.

Questions 1 through 3 refer to the Football Ticket Sales graph on page 146.

1. In what year did ticket sales stay the same as the year before?

2. In what year did ticket sales increase by about 1,000 tickets from the year before?

3. In which year were the fewest tickets sold?

Questions 4 through 7 refer to the information and graph below.

Aaron is taking trumpet lessons. To encourage him to practice, his mother is graphing the number of hours he practices each week.

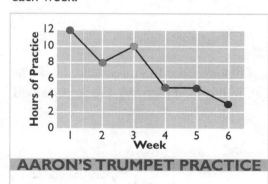

AARON'S TRUMPET PRACTICE

4. How many hours did Aaron practice during the fourth week?

5. In which week did Aaron practice more hours than he had the week before?

6. In which week did Aaron show the greatest decrease in practice hours from the week before?

7. Do you think it is more likely that Aaron will practice 2 hours or 10 hours in Week 7? Explain your thinking.

Questions 8 and 9 refer to the Elio's Pizza Sales graph on page 146.

8. In which month did the restaurant show a decrease in sales of Chicago-style pizzas?

9. Based on the graph, do you think the restaurant will always sell more Chicago-style pizzas than stuffed-crust pizzas? Explain your thinking.

Questions 10 through 13 refer to the following information and graph.

Three times a year, the Department of Water and Power sends its customers a graph of their water usage habits for four months. The graph sent to one family is shown below.

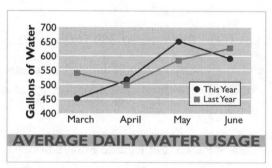

AVERAGE DAILY WATER USAGE

10. In which month of this year did the family use the greatest amount of water?

11. In which month of last year did the family show a decline in water usage from the month before?

12. In which month was last year's and this year's daily average most nearly the same?

13. In which month of this year was the family's daily average about 580 gallons?

Answers and explanations start on page 213.
For more practice with graphs, see page 185.

Using Circle Graphs

A **circle graph** uses a circle divided into sections to show how a whole amount is divided into parts. The size of the sections shows the relative size of the parts. For instance, suppose a company spends three times as much on payroll as it does on rent. On a graph of the company's expenses, the payroll section would be three times larger than the rent section.

The sections in a circle graph are most often labeled with percents. Since the circle represents the whole amount, the sections will total 100%. As you study the circle graphs in this lesson, think about the relationship between fractions and percents. A section labeled 50% will be $\frac{1}{2}$ of the whole circle. You can use your understanding of fractions to solve problems related to circle graphs.

EXAMPLE: A newspaper published this graph to show the results of a city election for a new mayor. If 3,000 people voted, how many voted for Perez?

Step 1. Find the facts. The section labeled Perez shows 46%. You need to find 46% of 3,000.

Step 2. Use proportion. $\frac{x}{3,000} = \frac{46}{100}$
3,000 × 46 = 138,000, and
138,000 ÷ 100 = 1,380

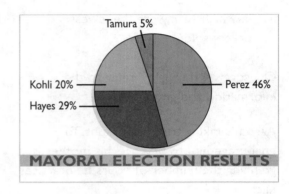

Tamura 5%
Kohli 20%
Hayes 29%
Perez 46%

MAYORAL ELECTION RESULTS

Answer: There were **1,380 votes** for Perez.

Circle graphs can also be labeled with amounts of money. Charities often send out circle graphs to show how each donation dollar is spent. In this kind of graph, the sections will total $1.

EXAMPLE: Pet Rescue Foundation depends on donations to rescue abandoned dogs and cats. The founder of the organization made this graph to show how donations are used. Out of a $50 donation, how much is spent on food?

Step 1. Find the facts. The section labeled Food says 20 cents. If 20 cents out of every dollar are spent on food, how much will be spent out of $50?

Step 2. Use proportion: $\frac{\$0.20}{\$1} = \frac{x}{\$50}$
$50 × $0.20 = $10, and $10 ÷ $1 = $10

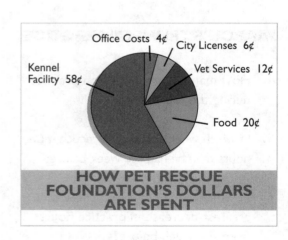

Office Costs 4¢
City Licenses 6¢
Kennel Facility 58¢
Vet Services 12¢
Food 20¢

HOW PET RESCUE FOUNDATION'S DOLLARS ARE SPENT

Answer: The amount spent on food will be **$10.**

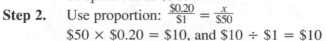

Solve as directed.

Questions 1 through 3 refer to the Mayoral Election Graph on page 148.

1. Which candidate received about $\frac{3}{10}$ of the vote?

2. What percent of the voters did <u>not</u> vote for Tamura? (*Hint:* Remember the whole circle represents 100% of the voters.)

3. What percent of the votes did the top two candidates receive?

Questions 4 through 7 refer to the graph below.

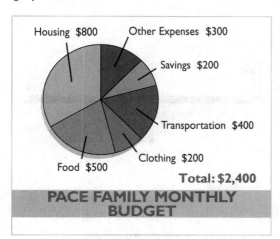

Housing $800
Other Expenses $300
Savings $200
Transportation $400
Clothing $200
Food $500
Total: $2,400

PACE FAMILY MONTHLY BUDGET

4. How much of the Pace family's monthly budget is left after the family pays its food expenses?

5. Which <u>two</u> items make up exactly 50% of the family's monthly budget?

6. What fraction of the graph is the section labeled Transportation?

7. How much money would the Paces need to make per month so that $\frac{1}{4}$ of their budget would be housing expense?

Questions 8 through 10 refer to the Pet Rescue Foundation graph on page 148.

8. Which <u>two</u> items in the graph use about $\frac{1}{3}$ of each donation dollar?

9. Sandra donated $200 to the Pet Rescue Foundation. How much of her donation will be spent on the kennel facility?

10. What percent of each donation is spent on office costs? (*Hint:* You need to find what percent 4 cents is of 1 dollar.)

Questions 11 through 14 refer to the following information and graph.

R&C Industries sent this graph to its employees.

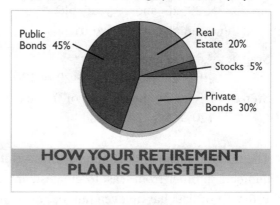

Public Bonds 45%
Real Estate 20%
Stocks 5%
Private Bonds 30%

HOW YOUR RETIREMENT PLAN IS INVESTED

11. If an employee has $500 in the retirement plan, how much is invested in stocks?

12. In what two items do employees invest $\frac{3}{4}$ of their retirement dollars?

13. Devon has $4,000 in his retirement plan. How much is invested in real estate?

14. What percent of an employee's retirement money is <u>not</u> invested in private bonds?

Answers and explanations start on page 213. For more practice with graphs, see page 185.

"Our phone bill may be different each month, but we watch it for several months, we'll get an idea of what we tend to spend, so we can plan a budget."

Special Topics in Data Analysis

Mean, Median, and Mode

We use the word *average* in many ways. If you are of average height, you know that some people are taller than you and some are shorter. If you have an average shoe size, you know your size is shared by many people.

In math, *average* has a special meaning. **Average** describes the center of a set of data. However, there are three different ways that we can describe the center. They are called mean, median, and mode.

The **mean** is the arithmetic average. To find the mean, add the numbers, and divide by the number of items in the set of data.

EXAMPLE: The ages of five friends are 20, 25, 29, 26, and 20 years old. Find the mean age of the five friends.

Step 1. Add the ages:
20 + 25 + 29 + 26 + 20 = 120
Step 2. Divide by the number of friends:
120 ÷ 5 = **24 years**

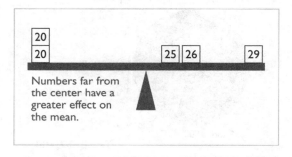

Numbers far from the center have a greater effect on the mean.

The **median** is the middle number in a set of data. To find the median, arrange the numbers in order from least to greatest, and find the value in the middle. [Special rule: If the set of data has an even number of items, there will be two numbers in the middle. The median is the mean (average) of these two numbers.]

EXAMPLE: Find the median age of the five friends.

Arrange the ages in order: 20 20 <u>25</u> 26 29. The median age is the middle value, **25.**

The **mode** is the number that occurs most often. Not all sets of data have a mode.

EXAMPLE: Find the mode of the ages of the five friends.

The age **20** occurs twice. The other ages occur only one time.

You now have three different ways to describe the average age of the group of friends. The mean (average) is 24, the median (middle age) is 25, and the mode (most common number) is 20. These three numbers give you more information about the group.

The problems on the GED Math Test will usually tell you whether to solve for mean, median, or mode. If a problem asks you to find the average and does not specify which kind, solve for the mean.

DATA ANALYSIS ▪ PRACTICE 5

Solve.

1. Pat drove from San Francisco to Nashville in five days. She drove the following distances:

   Day 1: 620 miles
   Day 2: 460 miles
   Day 3: 658 miles
   Day 4: 424 miles
   Day 5: 163 miles

   What was the mean number of miles she drove per day?

2. During June, Max went golfing five times. His scores during the month were 78, 74, 78, 109, and 76.
   a. What was his mean score?
   b. What was his median score?
   c. Which average, mean or median, best represents Max's normal score? Why?

3. During a weekend sale, the manager of a children's clothing store kept track of the sizes of clothes sold. She sold:

   Size 2T   45 items
   Size 3T   55 items
   Size 4T   30 items

   Which size represents the mode of the sizes sold?

4. The temperature at 3 P.M. for six days was 72°, 66°, 68°, 70°, 65°, and 74°. What was the median temperature for the six-day period?

5. The Mitchell family spent the following amounts on groceries over a 4-week period:

   Week 1 $118.50
   Week 2 $94.49
   Week 3 $108.12
   Week 4 $96.41

   What was the average amount the family spent on groceries per week?

Questions 6 through 8 refer to the table below.

Total Rainfall per Month Seattle, Washington	
Month	**Inches of Rainfall**
Oct.	3.4
Nov.	5.8
Dec.	6.0
Jan.	5.3
Feb.	4.0
Mar.	3.8
Apr.	2.5

6. For the months shown, what was the mean monthly rainfall in inches?

7. Which month had the median amount of rainfall?

8. If April's rainfall were changed from 2.5 to 0.5, which measure would change more: the mean or the median? Explain your thinking.

Answers and explanations start on page 213.

Probability

Probability is the study of chance. We use ratios or percents to express chance, or how likely it is that something will happen. Suppose a friend flips a coin and you call "heads." What is the chance that you will be right? You know that there are two possible outcomes when you flip a coin: heads or tails. If you call "heads," you have a 1 out of 2, or 50%, chance of being right.

You can find the probability of a simple random event by writing a ratio as a fraction. The denominator (bottom number) is the number of possible outcomes. The numerator (top number) is the number of **favorable outcomes,** or the event you are measuring.

Remember this formula: probability $= \frac{\text{favorable outcomes}}{\text{possible outcomes}}$

Study this example to see how to work with a more complex situation.

E X A M P L E : Stuart will win a board game if he rolls a 1 or a 2 on a regular 6-sided die. What is the probability that he will win the game on his turn?

Step 1. Find the possible and favorable outcomes. A regular die has six possible outcomes: 1, 2, 3, 4, 5, and 6. Two of the outcomes are favorable: either 1 or 2.

Step 2. Write the ratio and simplify: $\frac{\text{favorable outcomes}}{\text{possible outcomes}} = \frac{2}{6} = \frac{1}{3}$

Answer: There is a $\frac{1}{3}$, or **1 out of 3**, chance that Stuart will win on his turn.

On the GED Math Test, you may be asked probability questions about spinners, cards, or dice. You may also be asked to use the results of an experiment to find probability. **Experimental probability** is the ratio of the *favorable outcomes* to the *total number of trials*. A **trial** means doing the experiment a single time.

E X A M P L E : A bag holds 20 marbles. Some of the marbles are red, and some are white. Without looking, Maggie draws out a marble, records the color, and puts it back. After ten trials, she has these results:

white	red	white	white	white	white	red	white	red	white

Based on the experiment, what is the probability that the next marble Maggie chooses will be red? Express your answer as a percent.

These kinds of questions may seem tricky at first. There are two colors of marbles in the bag, but the probability isn't 1 out of 2 because there may be more of one color than the other in the bag. Based on the trials, **3 out of 10 samples** were red. Use this data to write the probability ratio.

$$\frac{\text{favorable outcomes}}{\text{possible outcomes}} = \frac{3 \text{ red marbles}}{10 \text{ trials}} = \frac{3}{10}$$

To change the ratio to a percent, write a proportion:

$$\frac{3}{10} = \frac{x}{100} \qquad 3 \times 100 = 300 \qquad 300 \div 10 = 30 \qquad \frac{3}{10} = 30\%$$

Answer: There is a **30% chance** of choosing a red marble.

Solve as directed. For questions 1 through 3, express probability as a ratio in lowest terms. For questions 4 through 6, express probability as a percent.

1. Owen is playing a card game with his sister. He has the following cards in his hand. If his sister chooses one card randomly, what is the chance that she will choose a card equal to or greater than 6?

2. Stephanie bought four raffle tickets for a contest at work. If 80 raffle tickets are sold in all, what is the chance that one of her tickets will be chosen?

3. The spinner below has eight equal sections.

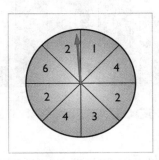

   If you spin the spinner, what is the probability that the pointer will land on an even number?

4. Jack is playing a game using a 20-sided die shown to the right. The die is numbered from 1 to 20. If he rolls the die once, what is the chance that he will get either 1, 2, 3, 4, or 5?

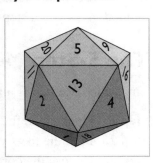

5. Suppose you have a deck of 50 cards with these symbols: ◆ ⊙ ▪ ★

   You perform an experiment to figure out how many of the cards in the deck have the ◆ symbol. Without looking, you choose a card and put it back in the deck. The results of 20 trials are shown below.

   a. Based on the trial, what is the probability that a card has the ◆ symbol?

   b. Based on the probability, how many of the 50 cards have a ◆ symbol?

6. At an emergency health clinic, 10 out of 25 employees are doctors. If one employee is chosen randomly, what is the chance that the worker will <u>not</u> be a doctor?

Answers and explanations start on page 213. For more practice with probability, see page 186.

The Problem-Solving Method

In the video, you saw how people collect and organize information in charts and graphs. This information is then used to solve problems.

A **problem-solving method** is a step-by-step strategy for solving word problems. Using a method consistently will help you be a better problem solver.

On the GED Math Test, most of the problems are word problems. The problem-solving method gives you a step-by-step way to approach word problems.

The 5-Step Method

The 5-Step Method is a way to get organized to solve word problems.

Step 1. Understand the question.
Step 2. Find the facts you need.
Step 3. Choose the correct operations.
Step 4. Set up and solve the problem.
Step 5. Check to make sure your answer is reasonable.

E X A M P L E : The bar graph below shows last week's attendance at River High School. What was the average daily attendance? Round your answer to the nearest whole number.

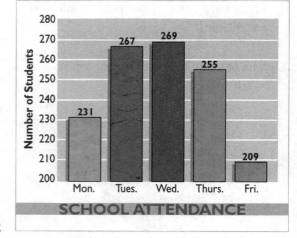

Step 1. *Understand the question.* The question asks you to find an average.

Step 2. *Find the facts you need.* The graph shows the daily attendance with each bar.

Step 3. *Choose the operations.* Add the attendance for all five days. Then divide by 5.

Step 4. *Set up and solve the problem.*
$$\frac{231 + 267 + 269 + 255 + 209}{5} = 246.2$$

Step 5. *Check to make sure your answer is reasonable.* The answer is reasonable because it is between the largest and smallest numbers for daily attendance.

Answer: Last week's average daily attendance at River High School was **246.**

THE PROBLEM-SOLVING METHOD

Use the 5-Step Method to solve each problem.

Questions 1 and 2 refer to the graph below.

Questions 3 and 4 refer to the graph below.

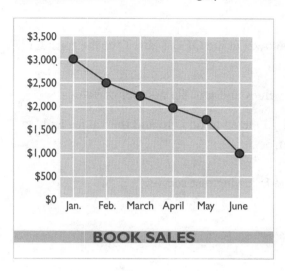

BOOK SALES

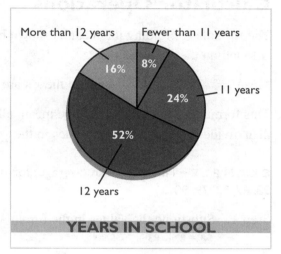

YEARS IN SCHOOL

1. Kevin recorded the amount of book sales for the first six months of the year. How much greater were book sales in January than in June?

   Question asks: _____

   _____

   Facts needed: _____

   Operation: _____

   Set up and solve: _____

   Check: _____

2. What were the average monthly book sales for the last two months?

   Question asks: _____

   _____

   Facts needed: _____

   Operation: _____

   Set up and solve: _____

   Check: _____

3. Javier made a graph of the number of years of education for each of the 50 employees at his hotel. How many employees have fewer than 11 years of education?

   Question asks: _____

   _____

   Facts needed: _____

   Operation: _____

   Set up and solve: _____

   Check: _____

4. What percent of the employees have 12 or more years of education?

   Question asks: _____

   _____

   Facts needed: _____

   Operation: _____

   Set up and solve: _____

   Check: _____

Answers and explanations start on page 213.

Data Analysis on the Calculator and Grid

You can use your calculator to find the mean, or average, of a set of numbers. You may also need to use your calculator to find the median, or middle value, of a set of numbers.

Calculator Operations

The formula for finding the mean is shown below. On the GED Math Test, the formula is found on the formulas page.

Mean $= \dfrac{x_1 + x_2 + \dots + x_n}{n}$ where the x's are values and n is the number of values

This formula shows that to find the mean, all you need to do is to add the values and then divide by the number of values in the set.

EXAMPLE: Find the mean (average) for the following test scores.
98, 87, 83, 79, 90

Step 1. Substitute the values in the formula.

$$\dfrac{98 + 87 + 83 + 79 + 90}{5}$$

Step 2. Add and divide.
$$\dfrac{437}{5} = \mathbf{87.4}$$

To use a calculator:

Step 1. Add the values and find the total.

98 $\boxed{+}$ 87 $\boxed{+}$ 83 $\boxed{+}$ 79 $\boxed{+}$ 90 $\boxed{=}$ 437

Step 2. Divide the total by the number of values.
437 $\boxed{\div}$ 5 $\boxed{=}$
The display shows $\boxed{\qquad 87.4\qquad}$

EXAMPLE: Find the median for the following race times (in seconds).
34.2, 34.5, 33.8, 33.9

Step 1. Arrange the numbers from smallest to largest.

33.8, 33.9, 34.2, 34.5

Step 2. Find the middle value. Since there is an even number of scores, the median is the average of the two middle numbers. Try this on your calculator:
Add the two middle values.
33.9 $\boxed{+}$ 34.2 $\boxed{=}$ 68.1
Divide the total by the number of values.
68.1 $\boxed{\div}$ 2 $\boxed{=}$ 34.05
The display shows $\boxed{\quad 34.05\quad}$

Notice that the median is a decimal number. Do not round the median.

Grid Basics

On the GED Math Test, you may need to fill in answers to data questions on a grid.

EXAMPLE: Last January in Marcie's hometown, the average amount of snowfall for four weeks was 8.6 inches. If the recorded snowfall amounts were 9.2, 8.0, and 7.9 inches for the first three weeks, how many inches of snow fell in the fourth week?

Step 1. First find the total amount of snowfall for the four weeks. Multiply the mean by 4, the number of values.

$8.6 \times 4 = 34.4$

Step 2. Subtract the sum of the three given snowfall amounts from the total.

$34.4 - (9.2 + 8.0 + 7.9) = 34.4 - 25.1 = \mathbf{9.3}$

GED MATH PRACTICE

DATA ANALYSIS ON THE CALCULATOR AND GRID

Use your calculator to solve each problem.

1. Find the mean:
   12, 8, 6, 15, 5

2. Find the median:
   23, 27, 19, 20, 15

3. Find the mean:
   3.2, 4.5, 6.0, 5.7

4. Find the median:
   $230, $450, $220, $300, $135, $600

5. Four homes sold for $75,000, $82,500, $77,500, and $95,000. What was the mean price for the homes?
   (1) $66,000
   (2) $80,000
   (3) $82,500
   (4) $89,500
   (5) $95,000

6. In three weeks, Tamra earned $237.00, $245.00, and $252.00. How much does she need to earn in the fourth week to average $250.00 per week?

Answers and explanations start on page 213.
For more practice with data analysis on the calculator and grids, see page 187.

GED Review: Data Analysis

Part I: Choose the <u>one best answer</u> to each question below. You <u>may</u> use your calculator.

<u>Questions 1 and 2</u> refer to the graph below.

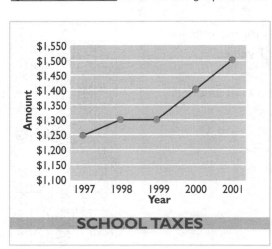

SCHOOL TAXES

<u>Questions 4 and 5</u> refer to the graph below.

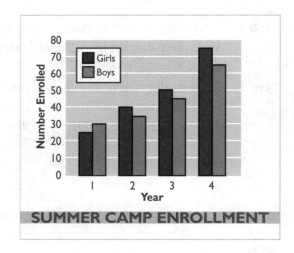

SUMMER CAMP ENROLLMENT

1. How much more did the Pirellis pay in 2001 than in 1997?
   (1) $250
   (2) $200
   (3) $150
   (4) $100
   (5) $50

2. On average, what did the Pirellis pay for school taxes over the five years?
   (1) $1,200
   (2) $1,300
   (3) $1,350
   (4) $1,400
   (5) $1,500

3. If Ty has 4 red socks and 6 white socks in a drawer, what is the probability that he will select a red sock at random?
   (1) 1/3
   (2) 1/2
   (3) 2/3
   (4) 2/5
   (5) 3/4

4. A new summer camp was built at Rainbow Lake. What was the total enrollment at the summer camp in the first year?
   (1) 25
   (2) 30
   (3) 40
   (4) 50
   (5) 55

5. In the fourth year, how many more girls than boys enrolled?

Part II: Choose the <u>one best answer</u> to each question below. You <u>may not</u> use your calculator.

<u>Questions 6 and 7</u> refer to the graph below.

<u>Questions 9 and 10</u> refer to the graph below.

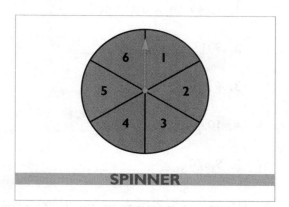

MONTHLY BUDGET

SPINNER

6. Lea made a graph of her monthly budget. What is the total amount of Lea's monthly budget?
   - **(1)** $1,350
   - **(2)** $1,250
   - **(3)** $1,100
   - **(4)** $800
   - **(5)** $500

7. What percent of her budget does Lea spend on rent?
   - **(1)** 8%
   - **(2)** 24%
   - **(3)** 30%
   - **(4)** 40%
   - **(5)** 47%

8. Joe recorded how long he walked each day for 5 days. The times are 50, 40, 65, 54, and 53 minutes. Find the mean number of minutes that he walked.
   - **(1)** 49.3
   - **(2)** 50
   - **(3)** 52.4
   - **(4)** 60
   - **(5)** 65.5

9. What is the probability of the spinner landing on an even number?
   - **(1)** $\frac{1}{6}$
   - **(2)** $\frac{1}{3}$
   - **(3)** $\frac{1}{2}$
   - **(4)** $\frac{2}{3}$
   - **(5)** $\frac{5}{6}$

10. What is the probability of the spinner landing on a number greater than 2?

①	①	①
⊙	⊙	⊙	⊙	⊙
⓪	⓪	⓪	⓪	⓪
①	①	①	①	①
②	②	②	②	②
③	③	③	③	③
④	④	④	④	④
⑤	⑤	⑤	⑤	⑤
⑥	⑥	⑥	⑥	⑥
⑦	⑦	⑦	⑦	⑦
⑧	⑧	⑧	⑧	⑧
⑨	⑨	⑨	⑨	⑨

Answers and explanations start on page 214.

Place Value, pages 26–27

A. Write the value of the underlined digit. The first problem is done for you.

1. 1,2<u>4</u>9   4 × 10 = 40 _____

2. 90<u>1</u> _____

3. <u>6</u>,288 _____

4. 1<u>0</u>,400 _____

5. 9,3<u>7</u>0 _____

6. 45,<u>6</u>72 _____

7. 3<u>9</u>9,000 _____

8. 7,<u>9</u>22,875 _____

9. 3<u>0</u>2,406 _____

10. <u>1</u>,950,000 _____

B. Write the number in words. The first problem is done for you.

11. 192   <u>one hundred ninety-two</u>

12. 347 _____

13. 4,006 _____

14. 7,230 _____

15. 34,851 _____

16. 70,500 _____

17. 323,000 _____

18. 560,700 _____

19. 4,200,000 _____

20. 7,000,401 _____

C. Write these numbers using digits. The first problem is done for you.

21. three hundred seventy-six <u>376</u>

22. two hundred eight _____

23. nine hundred thirty _____

24. eight thousand, twenty-four _____

25. one thousand, five _____

26. sixty thousand, nine hundred _____

27. ninety-five thousand, twenty-four _____

28. three hundred thousand, fifty _____

29. five million, two hundred thousand _____

30. one million, eight _____

Answers and explanations start on page 214.

Comparing and Ordering, pages 28–29

A. **Compare each pair of numbers. Write <, >, or = in the blank. The first problem is done for you.**

1. 89 __<__ 98

2. 370 _____ 307

3. 525 _____ 525

4. 6,020 _____ 6,002

5. 20,000 _____ 21,000

6. 1,005,300 _____ 1,050,300

7. 7,995 _____ 7,995

8. 81,582 _____ 81,580

9. 527 _____ 447

10. 205,566 _____ 204,966

11. 7,273,630 _____ 7,273,640

12. 4,301 _____ 4,103

B. **Write the numbers in order:**

from least (smallest) to greatest (largest)

13. 145   149   140

_____   _____   _____

14. 3,560   4,540   3,980

_____   _____   _____

from greatest (largest) to least (smallest)

15. 34,945   34,495   34,954

_____   _____   _____

16. 5,871   581   5,081

_____   _____   _____

C. **Solve.**

17. Janet got two bids to buy and install a new furnace. The first bid was $2,450, and the second bid was $2,540. Which is the lower bid?

Questions 18 through 20 refer to the table.

House	Selling Price
A	$95,800
B	$99,200
C	$98,900
D	$95,700

18. Which house sold for the greatest amount?

19. Which house sold for the least amount?

20. Write the selling prices in order from largest to smallest.

21. Write the names of the Great Lakes in order from <u>smallest to largest</u> size.

Great Lake	Area, sq. mi.
Erie	9,940
Huron	23,010
Michigan	22,400
Ontario	7,540
Superior	31,820

22. The numbers of registered voters for three counties are 25,061, 24,942, and 25,807. Write the numbers of voters in order from <u>greatest to least</u>.

23. JP Corporation had first-quarter revenues of $1,463,900 and $1,464,900 in revenues for the second quarter. Which quarter had the higher revenues?

Answers and explanations start on page 214.

Number Patterns, pages 30–33

A. For each pattern, fill in the missing number(s). The first problem is done for you.

1. 12 14 16 18 20 <u>22</u>

2. 37 33 29 25 21 _____

3. 105 95 85 75 65 _____

4. 9 18 _____ _____ 45 54

5. 30 35 _____ 45 50 _____

6. 150 130 110 90 70 _____

7. 40 48 56 _____ 72 _____

8. 925 825 725 625 525 _____

B. Answer each question.

9. Draw the figure that comes next in this pattern.

10. Draw the symbol that comes next in this pattern.

C. Solve.

11. By January, Joe lost a total of 6 pounds. By February, he lost a total of 12 pounds. He lost a total of 18 pounds by March and 24 pounds by April. If the pattern continues, how many total pounds will he lose by May?

12. Three years ago the population of a small town was 1,230. Two years ago the population was 1,240, and last year the population was 1,260. The population this year is 1,290. If the pattern continues, what will the population be next year?

13. Which of the following numbers are multiples of both 3 and 6?

    6 9 12 15 18 21 24 27

14. Jen gave her daughter 1 penny on Monday, 2 pennies on Tuesday, 4 pennies on Wednesday, and so on. If she continues to give her daughter pennies for a week using this pattern, how many pennies will Jen give her daughter on Sunday? *Hint:* Use multiplication to find this pattern.

    M   T   W   T   F   S   S
    1   2   4   8   16 32 ___

15. In May, Peter washed his car on the 3rd, 10th, 17th, and 24th. If the pattern continues, what is the next day in May that Peter will wash his car?

16. Which of the following numbers are multiples of both 2 and 7?

    14 21 28 35 42 49

Answers and explanations start on page 214.

Calculator Basics, pages 36–37

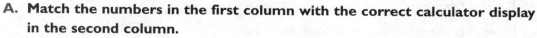

A. Match the numbers in the first column with the correct calculator display in the second column.

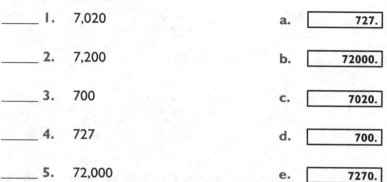

_____ 1.	7,020	a.	727.
_____ 2.	7,200	b.	72000.
_____ 3.	700	c.	7020.
_____ 4.	727	d.	700.
_____ 5.	72,000	e.	7270.
_____ 6.	70,200	f.	7200.
_____ 7.	7,270	g.	70200.

B. Practice entering the numbers below in your calculator. Clear your display before each entry.

8. 503

9. 9,437

10. 20,461

11. 610

12. 324,589

13. 6,598,200

14. 98,003,736

15. 800,700

C. Use the backspace key ▶ or the All Clear key AC on your calculator to make the corrections described below. (Note: If your calculator does not have these keys, use the Clear key C.)

16. Enter 246. Correct your entry to show 245.

17. Enter 5,090. Correct your entry to show 5,009.

18. Enter 6,672. Correct your entry to show 6,372.

19. Enter 84,987. Correct your entry to show 849,877.

20. Enter 35,060. Correct your entry to show 30,560.

Answers and explanations start on page 215.

Adding and Subtracting Whole Numbers, pages 42–45

A. Solve. The first problem is done for you.

1. $\begin{array}{r} 26 \\ + 43 \\ \hline 69 \end{array}$

3. $\begin{array}{r} 327 \\ + 51 \\ \hline \end{array}$

5. $\begin{array}{r} 6,895 \\ + 2,009 \\ \hline \end{array}$

7. $\begin{array}{r} 25 \\ 41 \\ + 638 \\ \hline \end{array}$

2. $\begin{array}{r} 97 \\ - 45 \\ \hline \end{array}$

4. $\begin{array}{r} 5,604 \\ - 3,725 \\ \hline \end{array}$

6. $\begin{array}{r} 4,050 \\ - 730 \\ \hline \end{array}$

8. $\begin{array}{r} 8,000 \\ - 933 \\ \hline \end{array}$

B. Rewrite the problems in a column and solve. The first problem is done for you.

9. $142 - 59$ $\begin{array}{r} 142 \\ - 59 \\ \hline 83 \end{array}$

11. $4,500 - 625$

10. $103 + 28 + 935$

12. $265 + 910 + 825$

C. Solve these problems using mental math.

13. $27 + 67$

15. $79 + 43$

17. $104 - 28$

14. $84 - 55$

16. $32 - 16$

18. $57 + 63$

D. Solve.

19. The Ski Club's expenses for its last weekend ski trip are shown below:

    Travel: $180
    Motel: $352
    Food: $256
    Skiing: $409

    What were the total expenses for the trip?

20. Jeff earned $25,800 working construction last year. His brother earned $27,100 painting houses. How much less did Jeff earn than his brother?

Questions 21 and 22 refer to the table below.

College Majors	Number of Students
Math	4,249
English	5,678
History	3,361
Science	2,432

21. How many more students have English than history as their major?

22. What is the total number of math and science majors?

Answers and explanations start on page 215.

Multiplying and Dividing Whole Numbers, pages 46–49

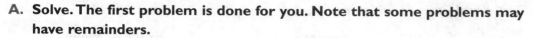

A. Solve. The first problem is done for you. Note that some problems may have remainders.

1. $\begin{array}{r} 23 \\ \times\ 9 \\ \hline 207 \end{array}$

4. $6\overline{)2{,}412}$

7. $\begin{array}{r} 502 \\ \times\ 16 \\ \hline \end{array}$

2. $8\overline{)176}$

5. $\begin{array}{r} 246 \\ \times\ 5 \\ \hline \end{array}$

8. $35\overline{)14{,}246}$

3. $\begin{array}{r} 185 \\ \times\ 7 \\ \hline \end{array}$

6. $18\overline{)1{,}368}$

9. $\begin{array}{r} 500 \\ \times\ 30 \\ \hline \end{array}$

B. Solve. The first problem is done for you.

10. $575 \div 23 = $ $23\overline{)575}^{\,25}$

13. $924 \times 50 = $

11. $32 \times 41 = $

14. $2{,}465 \div 42 = $

12. $500 \div 20 = $

15. $706 \times 300 = $

C. Solve.

16. Luke drives 27 miles round-trip to work each day. How far does he drive to and from work in 5 days?

17. Pam earns $1,780 per month. How much does she earn in a year?

18. Jesse paid a total of $4,500 last year for house payments. How much did he pay each month?

19. Ryan traveled 2,345 miles on his last sales trip. If his car gets 35 miles to the gallon, how many gallons of gasoline did he use on the trip?

20. Rosemary uses 1 gallon of juice each day in her day-care center. If each child drinks 8 ounces of juice, how many children are in her day-care center? (1 gallon = 128 ounces)

21. For a city in Florida, the newspaper headline read:

> **Record Rainfall This Year**
> *72 inches in 12 months*

What was the average rainfall per month?

Answers and explanations start on page 215.

Estimating, pages 50–53

A. Use rounding to estimate each answer. Then find the exact answer. The first problem is done for you.

		Estimate	**Exact Answer**
1.	2,345 + 12,892	2,300 + 13,000 = 15,300	15,237
2.	4,581 − 3,058		
3.	389 × 31		
4.	5,341 ÷ 12		

B. Show how you would solve the problem using front-end estimation. Then find the exact answer. The first problem is done for you.

		Estimate	**Exact Answer**
5.	$4,973 + $7,026	$4,000 + $7,000 = $11,000	$11,999
6.	523,980 − 293,487		
7.	1,878 × 39		
8.	21,895 ÷ 23		

C. Show how you would estimate the answer using compatible numbers. Then find the exact answer. The first problem is done for you.

		Estimate	**Exact Answer**
9.	405 ÷ 22	400 ÷ 20 = 20	18 r 9
10.	7,892 ÷ 39		
11.	9,856 ÷ 51		

D. Estimate an answer using whichever method you choose. Then find the exact answer.

12. Anna spent $72 on groceries, $45 on clothes, and $137 on car repairs. How much did she spend in all?

13. Vanessa needs $350 to pay tuition. If she has $259 in her savings account, how much more money does she need?

14. Spencer Theater has 11 rows of seats. If there are 18 seats in each row, how many seats does the theater have?

15. Joan gave $12,416 to her 4 children. If they each received an equal amount, how much was given to each child?

Answers and explanations start on page 215.

Calculator Operations and Grid Basics, pages 56–57

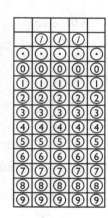 hidden

A. Use your calculator to solve the following problems. The first problem is done for you.

1. 670 + 304 = 974

2. 1,440 ÷ 60 =

3. 2,030 × 9 =

4. 207 − 188 =

5. 99 + 21 × 13 =

6. (800 − 15) ÷ 5 =

7. 786 + 1,340 =

8. 6,330 ÷ 15 =

9. (599 + 41) × 12 =

10. 4,500 − 892 =

B. Solve the problem. Fill in the grid. You **may** use your calculator when you think it will be helpful. For more information on grids, refer to page 222 in the Handbook.

11. Mary borrowed $480 for a new refrigerator. If she pays back the loan plus $48 in interest in 6 payments, how much will each payment be?

13. Jordon earns a total of $1,400 each month. He has $280 in deductions each month. How much does he take home in <u>one year</u>?

12. Carol drove 175 miles each day on a five-day trip. How many miles did she drive in all?

14. Jeff earned $3,500 on one construction job. On two more jobs, he earned $2,300 and $4,900. How much did he earn in all on the three jobs?

Answers and explanations start on page 215.

Rounding and Comparing Decimals, pages 64–65

A. Round as directed. The first problem is done for you. As a reminder, you may want to underline the place value to which you are rounding.

1. Round $1.4<u>9</u>9 to the nearest cent.
$1.50

2. Round 36.555 to the nearest tenth.

3. Round 9.3674 to the nearest hundredth.

4. Round 2.39 to the nearest whole number.

5. Round 27.615 to the nearest hundredth.

6. Round $13.78 to the nearest dollar.

7. Round $0.561 to the nearest cent.

8. Round 0.082 to the nearest tenth.

B. Compare each pair of numbers. Write <, >, or = in the blank. The first problem is done for you.

9. 29.3 __>__ 23.9

10. 6.75 _____ 6.73

11. 9.8 _____ 9.80

12. 0.251 _____ 0.151

13. 4.89 _____ 4.891

14. 0.18 _____ 0.099

15. 0.064 _____ 0.046

16. 2.315 _____ 3.415

17. 5.3 _____ 5.30

18. 7.23 _____ 7.233

C. Use rounding, comparing, and ordering to answer the following questions.

19. Arrange the following amounts from <u>greatest to least</u>: $31.25, $32.50, $31.55.

20. A bottle of molasses contains 882 milliliters. Round 882 milliliters to the nearest hundred milliliters.

21. The weights of 4 packages of nuts are shown below.

pecans	1.12 pounds
peanuts	1.35 pounds
almonds	0.98 pound
cashews	1.53 pounds

List the nuts, based on weight, from <u>lightest to heaviest</u>.

22. On a road trip, Ramona needed to buy gasoline. She noticed the price per gallon at four gas stations: $1.46, $1.43, $1.41, and $1.45. Which is the least expensive price per gallon?

23. One cubic foot of water weighs 62.425 pounds. Round this number to the nearest tenth.

24. Two measurements were reported as 2.7844 inches and 2.7861 inches. Rounded to the nearest hundredth, which measurement is larger?

Answers and explanations start on page 216.

Adding and Subtracting Decimals, pages 66–69

A. Solve. The first problem is done for you.

1.
```
   4.9
 + 3.2
 ─────
   8.1
```

2.
```
   1.8
 − 0.6
```

3.
```
   7.39
 − 4.69
```

4.
```
  $36.42
 + 13.58
```

5.
```
   4.654
 − 0.938
```

6.
```
   0.31
 + 5.807
```

B. Rewrite the problems in a column and solve. The first problem is done for you.

7. $0.7 + 4.09 + 5.362 =$
```
    0.7
    4.09
 +  5.362
 ────────
   10.152
```

8. $\$0.62 - \$0.58 =$

9. $0.802 + 8.2 + 0.28 =$

10. $25.5 - 6.752 =$

11. $0.09 + 4 + 6.537$

12. $40 - 11.49 =$

C. Solve.

13. Gina earned a commission of $456.50 her first week selling magazine subscriptions. The second week she earned $542.70. How much more did she earn the second week than the first?

14. The measurements of the four sides of a flower garden are shown below. What is the total distance around the garden?

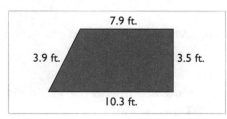

15. One new Brand A tire costs $75.89. One new Brand B tire costs $82.95. How much more does a Brand B tire cost than a Brand A tire?

16. Josh's dinner cost $12.50. The tax was $0.81, and he left $2.50 for the tip. How much did he spend in all?

17. At 6 A.M. the temperature was 21.3 degrees. By noon, the temperature had risen to 52.7 degrees. By how many degrees did the temperature rise?

18. Find the length of *x* in this figure.

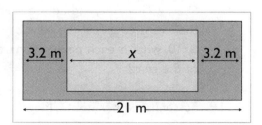

Answers and explanations start on page 216.

Multiplying and Dividing Decimals, pages 70–73

A. Solve. Round amounts of money to the nearest cent as needed. The first problem is done for you.

1.
$$\begin{array}{r} 5.7 \\ \times\ 6 \\ \hline 34.2 \end{array}$$

4. $0.6\overline{)\$4.95}$

7.
$$\begin{array}{r} 68 \\ \times\ 0.75 \\ \hline \end{array}$$

2. $5\overline{)21.5}$

5.
$$\begin{array}{r} \$8.40 \\ \times\ 3 \\ \hline \end{array}$$

8. $0.49\overline{)9.898}$

3.
$$\begin{array}{r} 0.04 \\ \times\ 0.9 \\ \hline \end{array}$$

6. $1.2\overline{)44.4}$

9.
$$\begin{array}{r} \$39.95 \\ \times\ 0.08 \\ \hline \end{array}$$

B. Solve. The first problem is done for you.

10. $3.96 \div 0.4 \quad 0.4\overline{)3.96}$ with quotient 9.9

13. $\$195.50 \times 0.075$

11. 0.007×1.2

14. $\$10 \div 24$

12. $\$5.40 \div 9$

15. 3.6×0.29

C. Solve. Round amounts of money to the nearest cent as needed.

16. Sierra Movie Theater sold 189 tickets for $6.50 each on Friday night. What was the total amount of ticket sales for that night?

17. Mountain Roasted coffee beans cost $7.49 per pound. How much will 1.5 pounds cost?

18. Sherry, Tina, Ana, and Louise split their rent equally. If the rent is $710, what is each person's share of the rent?

19. To refinance her house, Megan must pay a fee of 1.5 points or 0.015 times the amount of her new loan. If her new loan is $45,500, how much is the fee?

Questions 20 and 21 refer to the table and information below.

For tax purposes, Tim recorded his mileage on a business trip, the number of gallons of gas, and the cost of gas.

Miles	Gallons	Cost
235	9.4	$14.10

20. What was Tim's cost per gallon of gasoline?

21. How many miles per gallon did Tim's car get?

Answers and explanations start on page 216.

Decimal Calculator Operations and Grids, pages 76–77

A. Use your calculator to solve the following problems. The first problem is done for you.

1. $8.4 + 9.07 = \mathbf{17.47}$

2. $\$335 \div 4 =$

3. $6.3 \times 5.5 =$

4. $24 - 7.02 =$

5. $8.93 + 0.94 \times 7 =$

6. $14.3 - 7.03 =$

7. $27.09 \div 9 =$

8. $\$9.52 \times 0.08 =$

9. $\$10 - \$2.89 =$

10. $54.6 + 32.8 \div 4 =$

B. Solve each problem, and fill in the grid. You <u>may</u> use your calculator when you think it will be helpful. For more information on standard grids, refer to page 222 in the Handbook.

11. Myrna's car gets 35.5 miles per gallon. At that rate, how many miles could she travel on 15 gallons of gas?

13. The high temperatures for the last five days were 57.2, 59.0, 54.5, 60.1, and 49.7. How many degrees is the average temperature for the five days? (*Hint:* Find the average temperature by dividing the sum by the number of days.)

12. Carl packs 1 dozen identical parts in a box. If the parts weigh a total of 15 pounds, how many pounds does each part weigh?

14. Martina weighed 148.2 pounds two weeks ago. If she now weighs 145.8 pounds, how many pounds did she lose?

Answers and explanations start on page 216.

Fraction Basics, pages 82–83

A. Simplify these fractions. The first one is done for you.

1. $\dfrac{4}{8} = \dfrac{4 \div 4}{8 \div 4} = \dfrac{1}{2}$

3. $\dfrac{9}{36}$

5. $\dfrac{5}{30}$

2. $\dfrac{16}{18}$

4. $\dfrac{20}{50}$

6. $\dfrac{14}{56}$

B. Raise each fraction to higher terms shown by the given denominator. The first one is done for you.

7. $\dfrac{1}{2} = \dfrac{?}{12}$
   $\dfrac{1}{2} = \dfrac{1 \times 6}{2 \times 6} = \dfrac{6}{12}$

9. $\dfrac{3}{8} = \dfrac{?}{24}$

11. $\dfrac{4}{7} = \dfrac{?}{21}$

8. $\dfrac{2}{5} = \dfrac{?}{20}$

10. $\dfrac{4}{9} = \dfrac{?}{36}$

12. $\dfrac{6}{10} = \dfrac{?}{50}$

C. Compare each pair of fractions using >, <, or =. The first one is done for you.

13. $\dfrac{1}{2} > \dfrac{1}{4}$

   $\dfrac{1}{2} = \dfrac{2}{4}$

   $\dfrac{2}{4} > \dfrac{1}{4}$, so $\dfrac{1}{2} \ge \dfrac{1}{4}$

15. $\dfrac{1}{2} \underline{\quad} \dfrac{4}{8}$

17. $\dfrac{1}{3} \underline{\quad} \dfrac{5}{6}$

14. $\dfrac{3}{4} \underline{\quad} \dfrac{3}{8}$

16. $\dfrac{5}{8} \underline{\quad} \dfrac{10}{16}$

18. $\dfrac{3}{5} \underline{\quad} \dfrac{7}{10}$

D. Solve. Simplify your answer.

19. In a room containing 50 chairs, 35 are green. What fraction of the chairs is green?

20. The width of a field is 30 meters, and the length is 90 meters. The width is what fraction of the length?

21. Katy has $450 in her savings account. She withdrew $100. What fraction of her savings did she withdraw?

22. Samuel earns $40,000 per year. His son David earns $25,000 per year. David's salary is what fraction of his father's?

Answers and explanations start on page 216.

Adding and Subtracting Fractions, pages 84–89

A. Solve. Simplify if needed. The first one is done for you.

1. $\frac{1}{5}$
    $+\frac{3}{5}$
    ___
    $\frac{4}{5}$

7. $5\frac{2}{5}$
    $+9\frac{2}{5}$

13. $7\frac{7}{8}$
    $+\ \ \frac{3}{4}$

19. $6\frac{7}{10}$
    $+4\frac{2}{3}$

2. $\frac{7}{8}$
    $-\frac{3}{8}$

8. $10\frac{5}{6}$
    $-\ 3\frac{1}{6}$

14. $9\frac{3}{4}$
    $-\ \ \frac{3}{10}$

20. $1\frac{1}{3}$
    $-\ \ \frac{2}{5}$

3. $\frac{2}{7}$
    $+\frac{5}{7}$

9. $7\frac{3}{8}$
    $+6\frac{1}{8}$

15. $8\frac{2}{3}$
    $+\ \ \frac{4}{5}$

21. $15\frac{1}{4}$
    $+6\frac{7}{8}$

4. $\frac{7}{10}$
    $-\frac{3}{10}$

10. $18\frac{6}{7}$
    $-\ 9\frac{4}{7}$

16. $10\frac{5}{6}$
    $-\ \ \frac{3}{4}$

22. $11\frac{2}{5}$
    $-\ 4\frac{5}{6}$

5. $\frac{1}{3}$
    $+\frac{1}{4}$

11. $26\frac{1}{2}$
    $+15\frac{3}{10}$

17. 15
    $+\ \ \frac{2}{9}$

23. $32\frac{7}{8}$
    $+8\frac{5}{16}$

6. $\frac{5}{12}$
    $-\frac{1}{8}$

12. $9\frac{5}{8}$
    $-8\frac{1}{3}$

18. 5
    $-\ \frac{5}{12}$

24. $12\frac{1}{4}$
    $-\ 2\frac{5}{6}$

B. First decide whether to add or subtract. Then solve. Simplify if needed.

<u>Questions 25 and 26</u> refer to the table below.

Skirt Size	Yardage
12	$1\frac{3}{8}$
14	$2\frac{1}{4}$
16	3

25. How much more fabric will Susan need to make a size 16 skirt than a size 12 skirt?

26. How much fabric will Susan need to make two size 14 skirts?

27. Mai needs to work 35 hours each week. She has worked $25\frac{1}{2}$ hours. How many more hours does she need to work?

28. Jennifer plans to build a house on two adjacent lots. One lot is $1\frac{1}{2}$ acres, and the other is $\frac{3}{4}$ acre. What is the combined acreage of the two lots?

Answers and explanations start on page 217.

Multiplying and Dividing Fractions, pages 90–93

A. Solve. Simplify if needed. The first one is done for you.

1. $\frac{1}{2} \times \frac{1}{3} = \frac{1}{6}$

2. $\frac{3}{10} \div \frac{2}{5}$

3. $\frac{1}{6} \times \frac{3}{4}$

4. $\frac{9}{14} \div \frac{2}{7}$

5. $\frac{2}{3} \times \frac{2}{5}$

6. $\frac{2}{9} \times \frac{3}{8}$

7. $6 \div \frac{1}{6}$

8. $5 \times \frac{4}{9}$

9. $12\frac{1}{4} \div 2$

10. $1\frac{3}{11} \times 5$

11. $7 \div 1\frac{1}{5}$

12. $2\frac{3}{4} \div \frac{3}{4}$

13. $1\frac{1}{4} \times 3\frac{4}{5}$

14. $4\frac{1}{3} \div 1\frac{1}{3}$

15. $5\frac{1}{4} \times 2\frac{3}{7}$

16. $2\frac{9}{10} \div \frac{1}{5}$

17. $5\frac{1}{3} \times 1\frac{5}{6}$

18. $6\frac{1}{4} \times 5\frac{3}{5}$

B. First decide whether to multiply or divide. Then solve. Simplify your answers.

19. Michael earns $6 per hour working part-time at the library. If he works $4\frac{1}{3}$ hours each day, how much does he earn per day?

20. Rolando's car used $6\frac{1}{2}$ gallons of gasoline to travel 117 miles. At the same rate, how many miles could his car travel on 1 gallon of gasoline?

21. A candy recipe calls for $3\frac{3}{4}$ ounces of chocolate. How much chocolate is needed to triple the recipe?

22. Chef Lupe is serving a Monday night roast beef dinner special at his restaurant. If he serves $\frac{1}{3}$ pound per person, how many people can he serve from 3 roasts weighing a total of 26 pounds?

23. How many lengths of wire measuring $15\frac{3}{4}$ feet each can be cut from a spool with 315 feet of wire?

Questions 24 and 25 refer to the table below.

8-Foot Redwood Lumber	Actual Measurement
2 × 4	$1\frac{1}{2}$ in. thick × $3\frac{1}{2}$ in. wide
2 × 6	$1\frac{1}{2}$ in. thick × $5\frac{1}{2}$ in. wide

24. Rich is making a picnic table. The table will be 8 feet long and 33 inches wide. How many 8-foot boards will he need for the tabletop if he uses 2 × 6 boards?

25. Rich also needs to make 2 benches. The bench tops will be 8 feet long and 14 inches wide. How many 2 × 4 boards will he need for the 2 bench tops?

Answers and explanations start on page 217.

EXTRA PRACTICE

Fraction Calculator Operations and Grids, pages 96–97

A. Use your calculator to solve the following problems. The first problem is done for you.

1. $9\frac{2}{3} + 5\frac{1}{3} = 15$

2. $\frac{3}{5} \div \frac{1}{10} =$

3. $2\frac{3}{4} \times 7 =$

4. $6\frac{5}{8} - \frac{7}{8} =$

5. $\frac{1}{2} + \frac{3}{4} \times \frac{1}{3} =$

6. $10\frac{4}{9} - 3\frac{1}{3} =$

7. $12\frac{3}{8} \times \frac{1}{4} =$

8. $8\frac{7}{12} \div \frac{1}{6} =$

B. Solve the problem. Fill in the grid. You <u>may</u> use your calculator when you think it will be helpful. For more information on standard grids, refer to page 222 in the Handbook.

9. Michael took a 500-mile trip. On Monday, he drove $\frac{1}{2}$ of the distance. On Tuesday, he drove $\frac{4}{5}$ of the <u>remaining</u> distance. How many miles did Michael drive on Monday and Tuesday combined?

11. Erika earns $8 per hour. If she works $7\frac{1}{2}$ hours per day for 5 days, how many dollars will she earn?

10. To create a breakfast coffee blend, Paula combined $\frac{1}{3}$ pound of French Roast, $\frac{1}{2}$ pound of Kona, and $\frac{3}{4}$ pound of Hazelnut coffee. How many pounds of breakfast coffee blend does she have?

12. A recipe calls for $\frac{3}{4}$ cup of sugar. How many cups of sugar should be used if you double the recipe?

Answers and explanations start on page 217.

Ratios and Rates, pages 102–105

A. **Write each ratio as a fraction and simplify. The first problem is done for you.**

1. 24:4 $\frac{24}{4} = \frac{6}{1}$

2. 10 to 12

3. 5:9

4. 3 to 11

5. 14:5

6. 27 to 3

7. 52:26

8. 30 to 45

9. 10:100

B. **Find each unit rate. The first problem is done for you.**

10. 500 sheets of paper for $4
$\frac{500}{\$4} = \frac{125}{\$1}$ or
125 sheets per $1

11. 90 miles on 4.5 gallons of gas

12. $6.88 for 16 ounces

13. 200 square feet per 2 quarts of paint

14. 1.5 pounds per 6 cups of berries

15. 320 students on 8 buses

C. **Solve. Simplify your answers.**

16. A party punch is made using 3 liters of soda to 2 quarts of sherbet. What is the ratio of soda to sherbet?

17. Find the ratio of the width of the rectangle to the length of the rectangle.

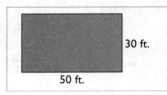

50 ft.
30 ft.

18. Jean earned $124 in 8 hours working as a carpenter. What is her pay rate in dollars per hour?

19. At a farmers' market, a bag of 12 oranges costs $3.84. What is the cost per orange?

20. Greenville's population is 12,500. Each day, 5,500 copies of the *Greenville News* are sold. What is the ratio of papers sold to the population of Greenville?

Questions 21 and 22 refer to the table below.

Westside School District	
School	**Students**
Adams High	250
Lincoln High	375
Washington High	150

21. What is the ratio of students at Washington High to students at Lincoln High?

22. What is the ratio of the number of students at Adams High to the total number of students in the three schools?

D. **Find the product that is the best buy (cheapest unit rate) in each advertisement.**

23.

Cereal
16 ounces for $2.40
24 ounces for $3.48

24.

Paper Towels
90 two-ply sheets per roll
Sparky—2 big rolls $2.45
Green Earth—3 big rolls $3.36
Cotton Clean—6 big rolls $5.88

Answers and explanations start on page 217.

MATH PRACTICE

Proportions, pages 106–109

A. Solve for the missing number in each proportion. The first problem is done for you.

1. $\frac{3}{4} = \frac{9}{x}$

    $3x = 4 \times 9$

    $x = \frac{4 \times 9}{3}$

    $x = 12$

5. $\frac{2}{3} = \frac{5}{x}$

9. $\frac{6}{7} = \frac{1.8}{x}$

2. $\frac{5}{6} = \frac{x}{18}$

6. $\frac{15}{10} = \frac{x}{2}$

10. $\frac{4}{9} = \frac{x}{4.5}$

3. $\frac{3}{10} = \frac{6}{x}$

7. $\frac{21}{30} = \frac{7}{x}$

11. $\frac{3}{8} = \frac{33}{x}$

4. $\frac{3}{8} = \frac{x}{24}$

8. $\frac{0.2}{0.5} = \frac{x}{20}$

12. $\frac{2}{5} = \frac{x}{2}$

B. Write a proportion and solve.

13. Rene scored 60 points in 4 basketball games. If she continues to score at the same rate, how many points will she score in 10 games?

14. Stevens College plans to have 20 computers for every 100 students. How many computers are needed for 1,500 students?

C. Solve.

15. A new video store is having a grand opening special.

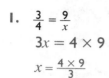

> **RENT**
> **New Releases**
> *Limited Time Only*
> **2 for $5.50**

At the same rate, how much would a customer pay to rent 6 videos?

16. On a blueprint, 3 inches = 9 feet. What is the length of a room that measures 5 inches on the blueprint?

17. The ratio of a liquid seaweed fertilizer concentrate mixed with water is 1:10. How many ounces of fertilizer are needed for 1 gallon of water? (*Hint:* 1 gallon = 128 ounces)

18. During a snowstorm, Capitan received 2 inches of snow in 3 hours. Snow continued to fall at the same rate until $7\frac{1}{2}$ inches of snow accumulated. How long did the snowstorm last?

Answers and explanations start on page 218.

Percents, pages 110–113

A. Write and solve a proportion for each problem. The first one is done for you.

1. What number is 30% of 42?

   $\frac{30}{100} = \frac{x}{42}$

   $100x = 30 \times 42$

   $x = \frac{30 \times 42}{100}$

   $x = $ **12.6**

2. $25 is what percent of $250?

3. 20% of what number is 8?

4. Find 60% of $12.

5. What percent of $10 is $7?

6. What is 75% of 50?

7. 6 is 5% of what number?

8. What percent is 13 of 52?

B. Solve.

9. Tami's dinner at the West Side Café was $8.00. She paid another 26% for tax and tip combined. How much did she spend on the tax and tip?

10. Out of a class of 30 students, 6 received an A. What percent received an A?

11. The population of Graniteville is 33,990. This is 110% of the population 5 years ago. What was the population 5 years ago?

12. The Cornerstone Bakery makes a profit of 17% on bread sales. What is the profit on bread sales totaling $690?

13. Lila spends $350 per month for rent. If she takes home $1,400 per month, what percent of her pay goes for rent?

14. Jake's class sold 125 tickets to the school play. This was 25% of all tickets sold. How many tickets were sold in all?

C. Solve these multi-step percent problems.

15. Phipps's Jewelry Store is discounting all diamond rings by 25%. What is the sale price of a ring that usually sells for $460?

16. High Country Furniture is going out of business. A sign in the window reads:

    > **Save 50% off**
    > **until December 1**

    After December 1, the owners decide to offer an additional 25% off all remaining merchandise. If the original price of a dining set was $890, how much will the dining set cost after December 1?

17. The price of coffee rose from $4.50 per pound to $5.40 per pound. What was the percent increase in the price of coffee?

18. A riding lawn mower with a list price of $799 is on sale for 15% off. What is the sale price of the mower?

19. Of $4,000 worth of tickets sold, $3,500 goes to charity. What percent did <u>not</u> go to charity?

Answers and explanations start on page 218.

MATH PRACTICE

Proportions and Percents with Calculators and Grids, pages 116–117

A. **Use your calculator to solve the following problems. The first one is done for you. For more information on percents, refer to page 223.**

1. 75 is what percent of 200?

   $\frac{75}{200} = 0.375$ or **37.5%**

2. What percent of 96 is 6?

3. $83 is 0.5% of what number?

4. What is 20% of 600?

5. 25 is 8% of what number?

6. 5.7 is 10% of what number?

7. What is $7\frac{1}{2}$% of $20?

8. 10 is 2.5% of what number?

9. $16 is what percent of $320?

10. What is 15% of $9.50?

B. **Solve each problem, and fill in the grid. You <u>may</u> use your calculator when you think it will be helpful. For more information on standard grids, refer to page 222 in the Handbook.**

11. The value of Ken's car has depreciated by 15% since he bought it a year ago. If he paid $12,000 for the car, how many dollars is it worth now?

13. Petra paid $90.00 for a new DVD player. If the sales tax rate was $7\frac{1}{2}$%, what was the total cost of the DVD player in dollars and cents?

12. Out of 1,400 employees, 22% took the early-retirement package offered. How many employees took early retirement?

14. A store sold 45% of its inventory of Cowboy brand jeans during a sale. If 36 pairs of jeans sold, how many pairs of jeans were in the original inventory?

Answers and explanations start on page 218.

Standard and Metric Measurement, pages 122–129

A. **Convert the measurements as directed. You may refer to the conversion tables on page 223 as needed. The first problem is done for you.**

1. How many feet are in 108 inches?

   $\frac{108}{12} = $ **9 ft.**

2. How many ounces are in $2\frac{1}{2}$ pounds?

3. How many cups equal 24 fluid ounces?

4. How many feet are in 0.5 yard?

5. How many tons equal 4,500 pounds?

6. How many minutes are in 300 seconds?

7. How many cups are in 1 quart?

8. How many meters are equal to 200 centimeters?

9. How many grams are in 1.2 kilograms?

10. How many liters are equal to 350 milliliters?

11. How many meters are in 0.2 kilometer?

12. How many grams are equal to 1,500 milligrams?

13. How many kiloliters are equal to 800 liters?

B. **Solve as directed. The first one is done for you.**

14. Add 2 lb. 8 oz. and 5 lb. 10 oz.

    $$
    \begin{array}{r}
    2\ \text{lb.} \quad 8\ \text{oz.} \\
    +\ 5\ \text{lb.} \quad 10\ \text{oz.} \\
    \hline
    7\ \text{lb.} \quad 18\ \text{oz.} = \textbf{8 lb. 2 oz.}
    \end{array}
    $$

15. Subtract 7 yd. 2 ft. from 10 yd.

16. Multiply 5 min. 24 sec. by 9.

17. Divide 3 gallons by 0.5 quart.

18. Find the sum of 700 m and 1,500 m. Express your answer in km.

19. How much heavier is 1g than 568 mg?

20. What is the product of 8.6 L multiplied by 7?

21. What is 23 kg divided by 4?

C. **Solve.**

22. Marcel works a 10-hour shift at the hospital. How much time is left in his shift if he has already worked 7 hours and 40 minutes?

23. Aleda used 3 feet 9 inches of fencing to make a trellis. How much does she have left if she originally had 10 feet of fencing?

24. The Green Herb Market buys tea in bulk in 1-kilogram packages. If the tea is repacked into 50-gram packages, how many packages can be made from 1 larger package?

25. A recipe calls for mixing white flour and whole-wheat flour in the ratio of 3:1. If Rafael uses 180 grams of white flour, how much whole-wheat flour will he need?

Answers and explanations start on page 218.

Perimeter, Area, and Volume, pages 130–131

A. Find the perimeter and area of each figure. The first problem is done for you.

1.

15 in.

$P = 4s$ $A = s^2$
$P = 4 \times 15$ $A = 15 \times 15$
$P = $ **60 inches** $A = $ **225 square inches**

2.

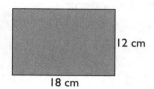

12 cm

18 cm

3.

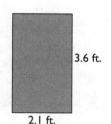

3.6 ft.

2.1 ft.

B. Find the volume of each box.

4.

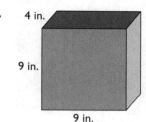

4 in.

9 in.

9 in.

5.

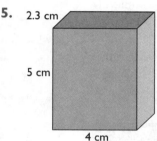

2.3 cm

5 cm

4 cm

6.

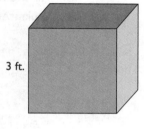

3 ft.

C. Solve.

7. Matt is putting weather stripping around a large picture window. The window is 6 feet wide and 4 feet tall. How many feet of weather stripping does he need?

8. A quilt measuring 4 feet by 6 feet is to be made out of 1-foot square blocks. How many quilt blocks will be needed?

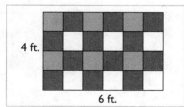

4 ft.

6 ft.

9. Leona is making a frame for a mirror that measures 30 inches by 18 inches. How many inches of framing material does she need?

10. A rectangular water tank measures 4 feet wide, 2.5 feet deep, and 8 feet long. What is the volume of the tank in cubic feet?

11. A post office box is 5 inches wide and 12 inches deep. If the height of the box is 7 inches, what is the volume of the box?

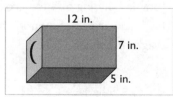

12 in.

7 in.

5 in.

12. Toni needs to paint his barn. Each of the four sides to be painted measures 24 feet long by 12 feet tall. How many square feet will he need to paint?

Answers and explanations start on page 218.

Irregular Figures, pages 132–133

Solve.

1. Cathy is building a planter for the corner of her porch. The drawing below shows the measurements of the planter.

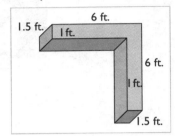

How many cubic feet of potting soil will she need to completely fill the planter?

<u>Questions 2 and 3</u> refer to the diagram below.

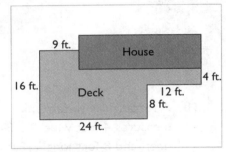

2. How many feet of railing are needed to go around the outside of the deck?

3. What is the area of the deck?

4. Joey is building a fence around his backyard as shown in the diagram below.

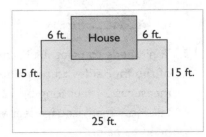

How many feet of fencing will he need to completely enclose the yard?

5. A 1-meter wide walkway is to be built around a swimming pool that is 10 meters by 16 meters. What is the area of the walkway?

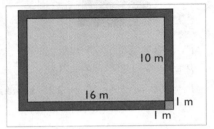

<u>Question 6</u> refers to the diagram below.

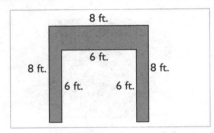

6. Sammi is replacing her kitchen countertop with ceramic tile. The diagram shows the measurements of each countertop. How many square feet of tile will she need?

7. A concrete footing is needed for each pier under a mountain cabin. Each footing will measure 12 inches by 12 inches and will be 18 inches deep. The layout of the piers is shown in the diagram below.

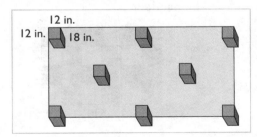

How many cubic feet of concrete will be needed to pour all of the footings?

Answers and explanations start on page 218.

GED Formulas, pages 136–137

A. Solve each problem below. You may refer to the formulas provided on page 224. You may use a calculator as needed. The first one is done for you.

1. Jackson drove for 2.5 hours at a speed of 70 miles per hour. How far did he drive?

    $d = rt$

    $d = 70 \times 2.5 =$ **175 miles**

2. A square sports field measures 300 feet on each side. What is the area of the field?

3. What is the area of the parallelogram?

    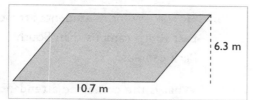

    6.3 m

    10.7 m

4. A large steel shipping crate measures 12 feet on each side and is 8 feet high. What is the volume of the crate?

5. A circular pond has a 5-meter diameter. What is the surface area of the water?

6. A triangular road sign measures 10 inches on each side. What is the perimeter of the road sign?

7. A cylindrical storage tank has the measurements shown in the figure below. What is the volume of the tank?

    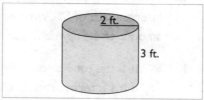

    2 ft.

    3 ft.

8. How much interest will Thomas pay if he obtains a 3-year loan of $1,500 and pays $7\frac{1}{4}$% simple interest per year?

B. Choose the expression that could be used to solve the problem.

9. A rectangular driveway measures 20 feet by 45 feet. Which expression could be used to find the area of the driveway?

    (1) 20×45
    (2) $2(20) + 2(45)$
    (3) $20 + 45$
    (4) $\frac{1}{2}(20 \times 45)$
    (5) $\frac{20 + 45}{2}$

10. Martha needs to sew trim around a pillow that measures 18 inches by 24 inches. Which expression could be used to find how many FEET of trim Martha needs?

    (1) $\frac{18 \times 24}{2}$
    (2) $2(18) + 2(24)$
    (3) $\frac{1}{2} \times 18 + 24$
    (4) 18×24
    (5) $\frac{(2)18 + (2)24}{12}$

11. Ami bought bags of popcorn for herself and her 4 children. If each bag cost $1.50, which expression shows the total cost?

    (1) $\$1.50 + 5$
    (2) $\$1.50 \div 5$
    (3) $\$1.50 \times 5$
    (4) $\$1.50 \times 4$
    (5) $\$1.50 + 4$

12. Which expression shows how many cubic feet of grain a silo with a diameter of 12 feet and a height of 15 feet will hold? (*Hint*: A silo is shaped like a cylinder.)

    (1) 3.14×6^2
    (2) $3.14 \times 2(6) \times 15$
    (3) $3.14 \times 12^2 \times 15$
    (4) $3.14 \times 6^2 \times 15$
    (5) 3.14×15

Answers and explanations start on page 219.

Tables, pages 142–143

Solve as directed. You may use a calculator.

<u>Questions 1 through 3</u> refer to the table.

Mortgage Table for a $100,000 Loan		
Interest Rate	Monthly Payment 30 Years	Monthly Payment 15 Years
6%	$600	$843
7%	$665	$898
8%	$733	$956

1. At an interest rate of 7%, how much more is the monthly payment for a 15-year mortgage than a 30-year mortgage?

2. With a 30-year mortgage and an interest rate of 8%, what is the total amount of the payments for one year?

3. What is the total amount Pete would pay if his interest rate is 6% and he chooses the 15-year mortgage?

<u>Questions 4 through 6</u> refer to the table.

World Population (in billions)		
	1985	2025 (projected)
Africa	560	1,495
Asia	2,819	4,759
Americas	666	1,035
Europe	770	863
Oceania	25	36
World	4,840	8,188

4. In 1985, the population of Africa was what percent of the world population? Round to the nearest whole percent.

5. What is the projected percent of increase for the world population (1985 to 2025)? Round to the nearest whole percent.

6. According to the projections in the table, which area of the world will have the greatest increase in population?

<u>Questions 7 through 9</u> refer to the pictograph.

Attendance at Community College Campuses
North Ridge 🧍🧍🧍🧍🧍 🧍🧍🧍🧍🧍 🧍🧍🧍
Rio Grande 🧍🧍🧍🧍🧍 🧍🧍🧍🧍🧍 🧍🧍🧍🧍🧍 🧍🧍
East Ridge 🧍🧍🧍🧍🧍 🧍🧍🧍
Pinnacle 🧍🧍🧍🧍🧍 🧍🧍🧍🧍🧍 🧍
South Park 🧍🧍🧍🧍🧍

Key: Each 🧍 represents 100 students.

7. How many students attend Rio Grande campus?

8. How many more students attend East Ridge campus than South Park campus?

9. What is the combined attendance for the two largest campuses? (*Hint:* You will need to add the symbols from two rows.)

<u>Questions 10 through 12</u> refer to the tax table.

Income Tax Rate of 15%	
Filing Status	Income Ranges
Single	$0–$24,650
Head of household	$0–$33,050
Married filing jointly	$0–$41,200
Married filing separately	$0–$20,600

10. Joe's filing status is single. If his income is $22,000, what is his income tax?

11. The Bransons are filing jointly. If their combined income is $38,000, what is their income tax?

12. For incomes higher than those in the table, the tax rate is higher. How much greater is the top income for a head of household than for the filing status of "single"?

Answers and explanations start on page 219.

Graphs, pages 144–149

Use the graphs to answer the questions.

<u>Questions 1 through 3</u> refer to the graph.

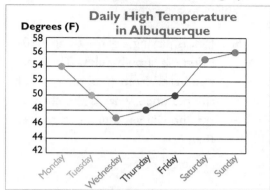

1. Which day had the highest temperature?

2. Which two days had the same temperature?

3. How much higher was the temperature on Monday than it was on Thursday?

<u>Questions 4 through 6</u> refer to the graph.

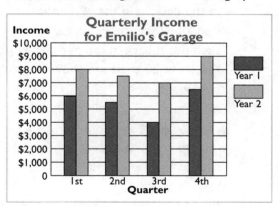

4. For which quarter was the increase in income the greatest over the two years?

5. What was the percent of increase from Year 1 to Year 2 for the 3rd quarter?

6. What was the total income for Year 2?

<u>Questions 7 through 9</u> refer to the graph.

7. For which two salespeople were the commissions in the ratio of 2:1?

8. What was the total amount of commission paid to the sales staff?

9. What percent of the total was George's commission? Round to the nearest whole percent.

<u>Questions 10 through 12</u> refer to the graph.

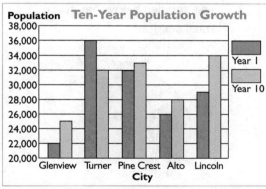

10. Which city had a decrease in population over the ten-year period shown?

11. What was the percent of increase for the city of Lincoln from Year 1 to Year 10? Round to the nearest whole percent.

12. For which city did the population increase by over 4,000?

Answers and explanations start on page 219.

Probability, pages 152–153

Solve as directed.

For <u>questions 1 through 4</u>, express probability as a ratio in lowest terms.

1. Paula is playing a card game with her father. She has the following cards in her hand. If her father chooses one card randomly, what is the chance that he will choose a card with a fish?

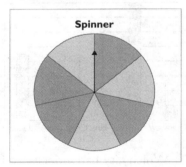

2. In one classroom, 12 out of the 28 students are girls. If one student is chosen randomly, what is the probability that the student will be a <u>boy</u>?

3. The spinner below has seven equal sections.

If you spin the spinner, what is the probability that the pointer will land on a blue section?

4. Tran has 20 socks in his drawer. Eight socks are white, 4 socks are blue, and 8 socks are brown. If he chooses one sock randomly, what is the probability that he gets a blue sock?

For <u>questions 5 through 7</u>, express probability as a percent.

5. Jack is playing a game using a 10-sided die shown below. The die is numbered from 1 to 10. If he rolls the die once, what is the chance that he will get a number greater than 3?

6. Suppose you have a deck of 40 cards with these symbols: ◆ ◉ ◼ ★

You perform an experiment to figure out how many of the cards in the deck have the ★ symbol. Randomly, you choose a card and then put it back in the deck. The results of 20 trials are shown below.

Based on the trials, what is the probability that a card has the ★ symbol?

7. There are 12 marbles in a box. Two marbles are black, 4 marbles are green, and 6 marbles are red. If one marble is chosen randomly, what is the chance that the marble will <u>not</u> be green?

Answers and explanations start on page 219.

Data Analysis with the Calculator and Grids, pages 156–157

A. Use your calculator to solve the following problems. Find the mean for each data set. If needed, round the answer to the nearest hundredth.

1. 2, 8, 9, 5, 5

2. 29, 32, 43, 30, 20, 48

3. 12.5, 14.7, 26.0, 15.1

4. $23.40, $4.50, $32.00, $86.30, $13.25, $6.80

B. Find the median for each data set. Do not round. You may use a calculator.

5. 7, 10, 5, 12, 15, 9

6. 213, 127, 99, 281, 315

7. 8.25, 6.57, 7.03, 4.76

8. $2,390, $4,540, $2,280, $3,000, $1,385, $1,600

C. Solve each problem, and fill in the grid. You <u>may</u> use your calculator. For more information on standard grids, refer to page 222 of the Handbook.

9. Maria received grades of 82, 79, 89, 91, and 88 on her math tests. What is the mean for her grades?

11. Five homes sold for $79,500, $83,300, $99,500, $70,000, and $105,900. What was the median price for the homes?

10. In four days, Carlos worked $6\frac{1}{2}$, 7, $5\frac{1}{4}$, and 8 hours. How many more hours does he need to work to average 7 hours per day for 5 days?

12. The annual salaries of six employees at Charter Communications are $34,000, $56,700, $49,250, $26,430, $39,100, and $87,400. What is the median salary for the six employees?

Answers and explanations start on page 219.

Math Posttest

Directions

The Math Posttest has 25 questions to measure your math and problem-solving skills. After you complete the test, check your answers on pages 198–199. Then use the evaluation chart on page 199 to identify the math skills that you still need to work on.

The Posttest consists of Part I and Part II. You will be allowed to use a calculator on Part I, but you are not required to use it. You <u>may not</u> use a calculator on Part II of the Posttest.

Most questions on the Math Posttest are multiple choice with five answer choices. A few questions do not have choices. For these questions, you should work the problem and then fill in your answer on the special answer grid on your answer sheet. Study the sample questions below to see how to use the answer sheet.

EXAMPLE: An 8-foot board is cut into three equal pieces. Disregarding any waste from the cuts, what is the length, in INCHES, of each piece?

(1) 24
(2) 28
(3) 30
(4) 32
(5) Not enough information is given.

① ② ③ ● ⑤

Answer: The correct answer is **(4) 32.** Fill in answer space 4 on the answer sheet.

EXAMPLE: Ken bought $2\frac{3}{4}$ pounds of apples and $1\frac{3}{4}$ pounds of peaches. How many pounds of fruit did he buy?

Answer: Ken bought **$4\frac{1}{2}$ pounds** of fruit. Examples of how the answer could be recorded correctly are shown below.

Remember these rules:

- Fill in only one circle in each column.
- You can start in any column as long as the answer fits within the grid.
- Mixed numbers such as $4\frac{1}{2}$ must be recorded as a decimal (4.5) or an improper fraction ($\frac{9}{2}$).

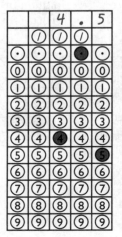

Posttest Answer Sheet

Part I

1. ① ② ③ ④ ⑤
2. ① ② ③ ④ ⑤
3. ① ② ③ ④ ⑤
4. ① ② ③ ④ ⑤
5. ① ② ③ ④ ⑤

6.

	/	/	/	
·	·	·	·	·
0	0	0	0	0
1	1	1	1	1
2	2	2	2	2
3	3	3	3	3
4	4	4	4	4
5	5	5	5	5
6	6	6	6	6
7	7	7	7	7
8	8	8	8	8
9	9	9	9	9

7. ① ② ③ ④ ⑤
8. ① ② ③ ④ ⑤
9. ① ② ③ ④ ⑤

10.

	/	/	/	
·	·	·	·	·
0	0	0	0	0
1	1	1	1	1
2	2	2	2	2
3	3	3	3	3
4	4	4	4	4
5	5	5	5	5
6	6	6	6	6
7	7	7	7	7
8	8	8	8	8
9	9	9	9	9

11. ① ② ③ ④ ⑤
12. ① ② ③ ④ ⑤
13. ① ② ③ ④ ⑤

Part II

14. ① ② ③ ④ ⑤
15. ① ② ③ ④ ⑤

16.

	/	/	/	
·	·	·	·	·
0	0	0	0	0
1	1	1	1	1
2	2	2	2	2
3	3	3	3	3
4	4	4	4	4
5	5	5	5	5
6	6	6	6	6
7	7	7	7	7
8	8	8	8	8
9	9	9	9	9

17. ① ② ③ ④ ⑤
18. ① ② ③ ④ ⑤
19. ① ② ③ ④ ⑤

20.

	/	/	/	
·	·	·	·	·
0	0	0	0	0
1	1	1	1	1
2	2	2	2	2
3	3	3	3	3
4	4	4	4	4
5	5	5	5	5
6	6	6	6	6
7	7	7	7	7
8	8	8	8	8
9	9	9	9	9

21. ① ② ③ ④ ⑤
22. ① ② ③ ④ ⑤
23. ① ② ③ ④ ⑤
24. ① ② ③ ④ ⑤
25. ① ② ③ ④ ⑤

Math Posttest

Part I

Write your answers on the answer sheet provided on page 189. You may use a calculator for any of the items in Part I, but some of the items may be solved more quickly without a calculator.

Some of the questions will require you to use a formula. The formulas you need are given on page 224. Not all of the formulas on that page will be needed.

Choose the <u>one best answer</u> to each question.

Question 1 refers to the following information.

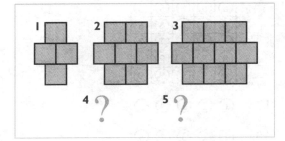

1. If the pattern continues, how many blocks would you need to build the FIFTH construction in this sequence?
   (1) 13
   (2) 16
   (3) 18
   (4) 20
   (5) 21

2. A machine can seal either 575 large packages or 850 small packages per hour. How many small packages can the machine seal in six hours?
   (1) 1,425
   (2) 1,650
   (3) 3,450
   (4) 5,100
   (5) 8,550

3. Frank pays for a software download service that costs $6 per month plus $3 for each download. Over a period of 4 months, he makes 18 downloads. What was Frank's total bill for the 4 months?
   (1) $36
   (2) $54
   (3) $78
   (4) $120
   (5) Not enough information is given.

Question 4 refers to the following information.

4. A recipe for buttermilk bread calls for the following liquid ingredients.

    $1\frac{1}{4}$ cups warm water

    $\frac{1}{3}$ cup vegetable oil

    $2\frac{1}{2}$ cups buttermilk

    How many cups of liquid ingredients are needed to make the bread?
    (1) $3\frac{1}{4}$
    (2) $3\frac{1}{3}$
    (3) $3\frac{1}{2}$
    (4) $3\frac{2}{3}$
    (5) $4\frac{1}{12}$

Question 5 refers to the following chart.

Art Store Price List	
Paper	**Price per Sheet**
Poster board	14.5 cents
Bristol board	22.5 cents
Foam board	54 cents

5. Not including tax, how many sheets of poster board could you buy for $15?
    (1) 27
    (2) 68
    (3) 100
    (4) 103
    (5) 217

6. In a large company, $\frac{5}{8}$ of the employees work full-time. Of those, $\frac{3}{10}$ are salespeople. What fraction of the company's employees are full-time salespeople? Express your answer in lowest terms.

    Mark your answer in the circles in the grid on your answer sheet.

Question 7 refers to the following information.

7. The regular price of the barbecue in the advertisement was $320. By what percent has the regular price been discounted?

   (1) 15%
   (2) 18%
   (3) 48%
   (4) 85%
   (5) Not enough information is given.

8. A parking lot has 420 regular spaces, 180 compact spaces, and 30 handicapped parking spaces. What is the ratio of regular to compact spaces?

   (1) 2:1
   (2) 3:2
   (3) 3:7
   (4) 7:3
   (5) 7:10

9. A grocery store is having a sale on frozen pizzas. The store is selling 5 pizzas for $8.50. Which of the following expressions could be used to find the cost of 12 pizzas?

   (1) $\$8.50 \times \frac{5}{12}$

   (2) $\$8.50 \div 12$

   (3) $\frac{12}{5 \times \$8.50}$

   (4) $\frac{5 \times 12}{\$8.50}$

   (5) $\frac{\$8.50 \times 12}{5}$

10. The rainfall totals over a 5-day period in Portland, Oregon, are shown below.

Day 1	1.44 cm
Day 2	1.3 cm
Day 3	2.26 cm
Day 4	2.6 cm
Day 5	0.5 cm

    What is the mean daily rainfall, in centimeters, for this period?

    Mark your answer in the circles in the grid on your answer sheet.

Questions 11 and 12 refer to the following diagram.

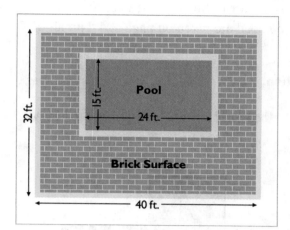

Question 13 refers to the following graph.

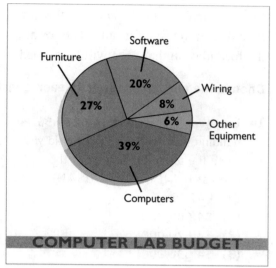

COMPUTER LAB BUDGET

11. A contractor plans to cover the shaded area around the pool with red brick. How many square feet does he plan to cover with brick?
- **(1)** 144
- **(2)** 360
- **(3)** 920
- **(4)** 1,280
- **(5)** 1,640

12. A blue tile border will cover the inside edge, or perimeter, of the 8-foot-deep pool. Which of the following expressions can be used to find the perimeter of the pool in feet?
- **(1)** 24 + 15 + 8
- **(2)** 24 + 24 + 15 + 15
- **(3)** 24 × 15 × 8
- **(4)** 40 + 40 + 32 + 32
- **(5)** 40 × 32

13. The graph shows the budget for a new computer lab for an elementary school. On which item will the school spend about one-fourth of its budget?
- **(1)** software
- **(2)** wiring
- **(3)** computers
- **(4)** furniture
- **(5)** other equipment

Part II

You may <u>not</u> use a calculator for the questions in Part II. Some of the questions will require you to use a formula. The formulas you need are given on page 224. Not all of the formulas on that page will be needed.

Choose the <u>one best answer</u> to each question.

14. Last year, a city spent $4,627,850 building new schools. How would you write the amount rounded to the nearest HUNDRED THOUSAND DOLLARS?
- **(1)** $4,600,000
- **(2)** $4,620,000
- **(3)** $4,630,000
- **(4)** $4,700,000
- **(5)** $5,000,000

15. Marcos sells water filtration systems to homeowners. He earns $150 per week plus $20 for each system he sells. Last week he worked for 32 hours. How much money did he earn?
- **(1)** $170
- **(2)** $202
- **(3)** $640
- **(4)** $790
- **(5)** Not enough information is given.

Question 16 refers to the following diagram.

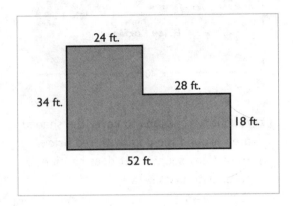

16. A neighborhood garden is formed from two rectangular plots of land. The measurements of the garden are shown above. How many feet of fencing will the neighbors need to enclose the garden?

Mark your answer in the circles in the grid on your answer sheet.

Question 17 refers to the following diagram of five books.

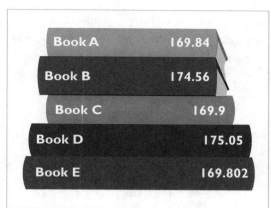

Book A	169.84
Book B	174.56
Book C	169.9
Book D	175.05
Book E	169.802

17. In a library, books are organized by call numbers written on the books. A shelf has the following sign:

> Call Numbers
> 169.841–174.5

Which of the books above belongs on this shelf?
(1) Book A
(2) Book B
(3) Book C
(4) Book D
(5) Book E

18. The shipping charges for a catalog order are based on the total weight of the order. Melinda is ordering the following items from a catalog.

Quantity	Item	Weight per item
2	sheet sets	3.6 pounds
4	bath towels	1.75 pounds
1	bedspread	2.5 pounds

Which of the following expressions could be used to find the total weight in pounds?
(1) $7 \times (3.6 + 1.75 + 2.5)$
(2) $(2 \times 3.6) + 2.5 + (4 \times 1.75)$
(3) $(4 \times 3.6) + (2 \times 1.75) + 2.5$
(4) $6 \times (3.6 + 1.75) + 2.5$
(5) $3.6 + 1.75 + 2.5$

19. A piece of plastic pipe is 16 feet long. How many pieces, each $\frac{2}{3}$ foot in length, could be cut from the pipe?
(1) 10
(2) 18
(3) 20
(4) 24
(5) 48

20. The ratio of wins to losses for a basketball team is 5:3. If the team has played 72 games, how many has the team won?

Mark your answer in the circles in the grid on your answer sheet.

21. Noah figures out that he spends 18% of his take-home pay on groceries for his family and 6% eating out. Which of the following expressions could be used to find the percent of his take-home pay that is NOT spent on food?

(1) 18% + 6%

(2) 18% − 6%

(3) 100% − (18% + 6%)

(4) 100% + (18% − 6%)

(5) 100% − (18% − 6%)

Question 22 refers to the following information and graph.

Deon made the following graph to show the number of hours he works each day during the week.

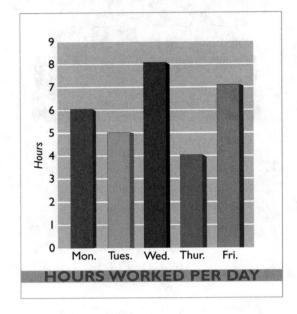

HOURS WORKED PER DAY

22. What is the ratio of hours Deon works on Monday to the total number of hours Deon works during the week?

(1) 1 to 6

(2) 1 to 5

(3) 1 to 4

(4) 3 to 20

(5) Not enough information is given.

Question 23 refers to the following graph.

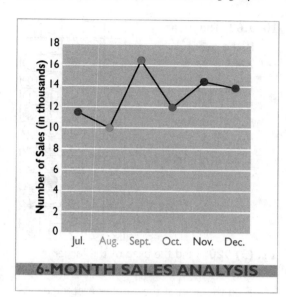

6-MONTH SALES ANALYSIS

23. In which month did the company have the greatest decrease in sales from the month before?
  (1) August
  (2) September
  (3) October
  (4) November
  (5) December

24. Jan ran eight laps on a 750-meter track. How many KILOMETERS did she run?
  (1) 0.6
  (2) 6
  (3) 60
  (4) 600
  (5) 6,000

25. A package weighs 52 ounces. Which of the following expressions could be used to change the weight to pounds?
  (1) 52×8
  (2) $52 \div 8$
  (3) $52 \div 10$
  (4) 52×16
  (5) $52 \div 16$

Answers and explanations begin on page 198.

Math Posttest Answer Key

Part I

1. **(2) 16** Each new picture adds three blocks. The third picture has 10 blocks. The fourth picture will have $10 + 3 = 13$ blocks, and the fifth will have $13 + 3 = 16$ blocks.

2. **(4) 5,100** $850 \times 6 = 5,100$ small packages. You do not need to know the number of large packages the machine can seal.

3. **(3) $78** (4 months $\times$ $6) + (18 downloads $\times$ $3) = $24 + $54 = $78

4. **(5) $4\frac{1}{12}$** Add:
$1\frac{1}{4} + \frac{1}{3} + 2\frac{1}{2} = 1\frac{3}{12} + \frac{4}{12} + 2\frac{6}{12} = 3\frac{13}{12} = 4\frac{1}{12}$

5. **(4) 103** 14.5 cents $= 0.145, and $15 \div $0.145 = 103$ plus a remainder. Disregard the remainder or round down, since you cannot buy part of a piece of poster board.

6. **$\frac{3}{16}$** Multiply.
$\frac{1}{\cancel{2}^{1}}{\cancel{8}} \times \frac{3}{\cancel{10}_{2}} = \frac{3}{16}$

Grid answer: 3 / 1 6

7. **(1) 15%** Subtract to find the difference between the regular price and the sale price: $320 - $272 = 48. Find the percent of decrease: $\frac{$48}{$320} = \frac{x}{100}$; $4,800 \div 320 = 15$, and $\frac{15}{100} = 15\%$.

8. **(4) 7:3** Compare 420 (the regular spaces) to 180 (the compact spaces). $\frac{420}{180} = \frac{7}{3}$

9. **(5) $\frac{$8.50 \times 12}{5}$** Set up the proportion. $\frac{5\text{ pizzas}}{$8.50} = \frac{12\text{ pizzas}}{x}$ To solve the proportion, multiply 8.50×12 and divide by 5. Only option (5) shows this series of operations.

10. **1.62** The mean is the average. Add: $1.44 + 1.3 + 2.26 + 2.6 + 0.5 = 8.1$. Divide by 5, the number of days: $8.1 \div 5 = 1.62$ centimeters.

Grid answer: 1 . 6 2

11. **(3) 920** Find the area of the larger rectangle: $40 \times 32 = 1,280$ square feet. Find the area of the smaller rectangle (the pool): $24 \times 15 = 360$ square feet. Subtract: $1,280 - 360 = 920$ square feet.

12. **(2) 24 + 24 + 15 + 15** The length of the pool is 24 feet, and the width is 15 feet. To find the perimeter, add the 4 sides.

13. **(4) furniture** One-fourth equals 25%. The budget item that is nearest 25% is furniture. You can also see that the size of the furniture section is closest to one-fourth of the entire circle.

Part II

14. **(1) $4,600,000** Look at the hundred thousands place: $4,627,850$. The number to the right, 2, is less than 5. Round down.

15. **(5) Not enough information is given.** You need to know the number of filtration systems he sold, not the number of hours he worked.

16. **172** The length of the missing side is the difference between 34, the opposite side, and 18: $34 - 18 = 16$ feet. Add to find the perimeter: $34 + 24 + 16 + 28 + 18 + 52 = 172$ feet.

Grid answer: 1 7 2

17. **(3) Book C** The number 169.9 is greater than 169.841 and less than 174.5.

18. **(2) (2 × 3.6) + 2.5 + (4 × 1.75)** For each item, you need to multiply the quantity by the weight. Then add the results. Only option (4) shows this series of operations.

19. **(4) 24** Divide: $16 \div \frac{2}{3} = \frac{16}{1} \times \frac{3}{2} = \frac{24}{1} = 24$

20. **45** The ratio of wins to total games played is 5 to 8 (5 wins + 3 losses = 8 games played). Write a proportion: $\frac{5}{8} = \frac{x}{72}$; $5 \times 72 = 360$, and $360 \div 8 = 45$

21. **(3) 100% − (18% + 6%)** If the total percent spent on food is the sum of 18% and 6%, the percent *not* spent on food is this sum subtracted from 100%.

22. **(2) 1 to 5** Add the values of the bars to find the total hours worked in the week: $6 + 5 + 8 + 4 + 7 = 30$ hours. Write the ratio of Monday's hours to the total, and simplify: $\frac{6}{30} = \frac{1}{5}$.

23. **(3) October** A decrease in sales is shown by the line graph sloping down from left to right. The greatest decrease is from September to October, a decrease of about 4,000 sales.

24. **(2) 6** Jan ran $750 \times 8 = 6{,}000$ meters. Since there are 1,000 meters in 1 kilometer, divide by 1,000. $6{,}000 \div 1{,}000 = 6$ kilometers

25. **(5) 52 ÷ 16** There are 16 ounces in one pound, so divide the number of ounces by 16 to change to pounds.

Evaluation Chart for Math Posttest

This evaluation chart will help you find the strengths and weaknesses in your understanding of math.

Follow these steps:
- Check your answers using the Answers and Explanations on pages 198–199.
- Circle the questions you answered correctly.
- Total your correct answers for each row.
- Use the results to see which math areas you need to study. If you got more than two wrong in any section, review the programs and workbook pages listed.

Questions	Total Correct	Program
1, 2, 3, 14, 15, 16	_____ / 6	20: Number Sense (pp. 20–39) 21: Problem Solving (pp. 40–59)
5, 17, 18	_____ / 3	22: Decimals (pp. 60–79)
4, 6, 19	_____ / 3	23: Fractions (pp. 80–99)
7, 8, 9, 20, 21, 22	_____ / 6	24: Ratio, Proportion, and Percent (pp. 100–119)
11, 12, 24, 25	_____ / 4	25: Measurement (pp. 120–139)
10, 13, 23	_____ / 3	26: Data Analysis (pp. 140–159)
TOTAL	_____ / 25	

Answers and Explanations

PROGRAM 20: NUMBER SENSE

Practice 1 (page 23)

All choices are correct. Your answers may vary.
Sample answers are given below.

1. figuring the tip at a restaurant
   buying groceries
   putting gas in the car
2. **d.** I find the address on a map and then make a list of directions to follow.
3. **d.** I would figure out how much interest the store is charging me and see whether I could get lower interest somewhere else.
4. **d.** I would buy the bigger bottle because my family will use it in time and it is a better buy.

Practice 2 (page 25)

All choices are correct. Your answers may vary.
Sample answers are given below.

1. **d.** I would read the tag on the sofa to find out its size. Then I would go home and measure that size along the wall to see if it would fit.
2. **d.** I would run the errand as soon as possible to make sure I have more than one hour to do my homework, just in case I need more time.
3. **d.** I would use objects to represent the three places.
4. **d.** I would estimate my family's usual medical expenses and then see which plan would save me the most money on those bills.

Practice 3 (page 27)

A.
1. $2 \times 1 = 2$
2. $5 \times 100 = \mathbf{500}$
3. $4 \times 10 = \mathbf{40}$
4. $3 \times 1,000 = \mathbf{3,000}$
5. $8 \times 100 = \mathbf{800}$
6. $7 \times 10,000 = \mathbf{70,000}$
7. $4 \times 1,000 = \mathbf{4,000}$
8. $3 \times 100,000 = \mathbf{300,000}$
9. $5 \times 1,000,000 = \mathbf{5,000,000}$
10. $6 \times 100 = \mathbf{600}$

B.
11. five hundred fifty-eight
12. two thousand, eight hundred three
13. twelve thousand, five hundred sixty
14. three hundred thousand, six hundred ninety-two
15. two million, five hundred seven thousand
16. five million, eight thousand, nine hundred fifty

C.
17. 336
18. 4,107
19. 24,000
20. 170,000
21. 15,098
22. 6,040,009

Practice 4 (page 29)

A.
1. $590 > 490$
2. $3,068 < 3,608$
3. $184 = 184$
4. $305,060 > 304,060$
5. $1,200,000 < 2,100,000$
6. $1,004 < 1,040$
7. $74,095 > 73,995$
8. $101,500 = 101,500$
9. $327 < 347$
10. $25,500 > 24,900$
11. $4,963,450 < 4,963,550$
12. $7,604 > 7,406$

B.
13. Dodgers
14. Food Saver
15. Greenscape
16. Maywood, Clayton, Greenview, Baker
17. $2,827; $2,158; $1,872
18. RS Industries

Practice 5 (page 31)

A.
1. **11** This is a series of odd numbers. Add 2 to continue the pattern.
2. **2** The rule for the pattern is "subtract 3."
3. **90** Count by 10s. The rule is "add 10" or "multiples of 10."
4. **56** The rule is "multiples of 8." You could also add 8 each time.
5. **10** The rule is "subtract 4."
6. **17** Increase the amount you add each time by 1. Add 1, then 2, then 3, and so on.
7. **54** The rule is "multiples of 6." You could also add 6 each time.
8. **650** The rule is "add 100."

B.
9. **15** Put a row of five blocks under the last figure.
10. **star** After each plus sign, another group of stars begins.

C.
11. **$55** The rule is "add $10."
12. **20** The rule is "multiples of 4." You could also add 4 each time.
13. **2:40 P.M.** The rule is "add 20 minutes."
14. **S** The letters stand for the days of the week. The missing day is Sunday.

Practice 6 (page 33)

A.
1. You should have shaded 9, 18, 27, 36, 45, 54, 63, 72, 81, 90, and 99.
2. The shaded boxes form a diagonal line that goes downward from right to left. You may also have noticed that the sum of the digits in each multiple of 9 is also a multiple of 9. Example: $1 + 8, 2 + 7, 3 + 6, 9 + 9$, and so on.
3. 108, 117, 126
4. **Multiples of 4:** 4, 8, 12, 16, 20, 24, 28, 32, 36, 40, 44, 48, 52, 56, 60, 64, 68, 72, 76, 80, 84, 88, 92, 96, 100
   Multiples of 8: 8, 16, 24, 32, 40, 48, 56, 64, 72, 80, 88, 96
5. Every multiple of 8 is also a multiple of 4.
6. All multiples of 8 are even. Since 153 is an odd number, it cannot be a multiple of 8.

B.
7. 21, 33
8. 28, 35
9. 36, 60
10. 110, 121

C. 11. 15, 30, 45

12. False. Of the numbers less than 100 that end in 8, only 8, 48, and 88 are multiples of 8.

GED Math Practice • Understand Math Problems (page 35)

1. (2) How long does it take Alex to walk three miles? Even though there is enough information to find how many miles Alex walks in a week, the question focuses on time.

2. (1) How much money does Carla make each month? Even though the problem mentions weeks, the question clearly asks about her monthly earnings.

3. (2) How far does Tanya travel to and from work each week? The question asks about each week.

4. (4) student, sunshine, basics, tropical In order from greatest to least, the weights are 40, 35, 30, and 25 pounds.

5. (3) 45 The shipment weights increase by 5 pounds each time. The next shipment weight after 40 pounds would be 45 pounds.

GED Math Practice • Calculator Basics (page 37)

A. Remember, the calculator display <u>does not</u> show commas. The display <u>does</u> show a decimal point in each number.

1. c
2. e
3. a
4. b
5. d

B. For larger numbers, enter only two or three digits at a time. Check the display after you enter each group of numbers.

6.	489.	9.	115987.
7.	6256.	10.	3468920.
8.	10573.	11.	23546989.

C. The corrected entries are shown below.

12.	223.	15.	783123.
13.	1015.	16.	20500.
14.	9968.		

GED Math Review (pages 38–39)

1. (5) 102,056

2. (4) 220,046

3. (5) 60,000 The 6 is in the ten thousands column, so its value is $6 \times 10,000 = 60,000$.

4. (3) $141,006 and $140,964 Compare the amounts in the revenue column. Remember to work from left to right when comparing place-value columns.

5. (4) Year 2 and Year 3 The expenses for both Years 2 and 3 are close to $81,000.

6. (2) Year 1, Year 4, Year 2, Year 3, Year 5 Find the expense column, and compare the values.

7. (5) 10, 8, 6, 4 The numbers are listed in order, starting with the heaviest.

8. (4) 70,000 The 7 is in the ten thousands column. Therefore, the 7 in 175,261 has a value of $7 \times 10,000 = 70,000$.

9. (3) 14,070 Fourteen thousand seventy means there are 14 thousands, 14,000, and 7 tens, 70, or 14,070.

10. (2) 6,300 > 6,030 Compare the place-value columns from left to right. Since both numbers have a 6 in the thousands column, compare the hundreds column. Since 3 is greater than 0, 6,300 is greater than 6,030.

11. (1) 6:35 Hitting the snooze button adds 5 minutes each time. If Marcus hits the snooze button 4 times, he adds 20 minutes to his original alarm time. You could also count by 5s to find the new time; start at 6:15 and count 6:20, 6:25, 6:30, and 6:35.

12. (4) 80 The pattern shows multiples of 8. In other words, add 8 to the last number.

13. (2) 14 Look for a pattern. Jean has been increasing the number of laps by 3 each week. By Week 4, she will be swimming 14 laps each time she swims.

PROGRAM 21: PROBLEM SOLVING

Practice 1 (page 43)

1.	59	10.	3,768
2.	196	11.	240
3.	557	12.	1,693
4.	3,673	13.	85
5.	5,295	14.	142
6.	3,787	15.	159
7.	233	16.	114
8.	1,256	17.	50
9.	554	18.	129

19. 361 points $78 + 92 + 86 + 105 = 361$

20. $2,395 $1,568 + $827 = $2,395$

21. $1,418 $1,199 + $219 = $1,418$

22. $1,544 $1,385 + $159 = $1,544$

Practice 2 (page 45)

1. 36 Check: $36 + 23 = 59$

2. 630 Check: $630 + 156 = 786$

3. 16 Check: $16 + 46 = 62$

4. 173 Check: $173 + 175 = 348$

5. 1,841 Check: $1,841 + 2,197 = 4,038$

6. 326 Check: $326 + 274 = 600$

7. 745 Check: $745 + 179 = 924$

8. 2,107 Check: $2,107 + 3,275 = 5,382$

9. 473 Check: $473 + 89 = 562$

10. 803 Check: $803 + 148 = 951$

11. 52 Check: $52 + 38 = 90$

12. 811 Check: $811 + 194 = 1,005$

13. 82

14. 48

15. 61

16. 49

17. 42

18. 38

19. 1,883 people $3,475 - 1,592 = 1,883$

20. 1,160 people $3,475 - 2,315 = 1,160$

21. 282 miles $600 - 318 = 282$

22. $694 $2,620 - $1,926 = 694

23. $2,169 $2,564 - $395 = $2,169$

Practice 3 (page 47)

A.
1. 128
2. 225
3. 360
4. 1,176
5. 1,544

6. 3,620
7. 3,430
8. 20,532
9. 10,920

B.
10. 6,150
11. 1,404
12. 62,400

13. 27,060
14. 38,346
15. 125,000

C.
16. **$4,140** $345 × 12 = $4,140
17. **3,600 labels**   120 × 30 = 3,600
18. **120 pictures**   24 × 5 = 120
19. **$504**   42 × $12 = $504
20. **1,785 calories**   85 × 3 × 7 = 1,785
21. **435 calories**   145 × 3 = 435

Practice 4 (page 49)

A.
1. **162**   162 × 5 = 810
2. **63**   63 × 9 = 567
3. **316**   316 × 7 = 2,212
4. **283**   283 × 13 = 3,679
5. **71**   71 × 19 = 1,349
6. **504**   504 × 28 = 14,112
7. **153 r3**   153 × 4 = 612; 612 + 3 = 615
8. **492 r9**   492 × 12 = 5,904; 5,904 + 9 = 5,913
9. **607**   607 × 32 = 19,424

B.
10. **842 r4**   842 × 6 = 5,052; 5,052 + 4 = 5,056
11. **96 r2**   96 × 23 = 2,208; 2,208 + 2 = 2,210
12. **63**   63 × 27 = 1,701
13. **158 r4**   158 × 34 = 5,372; 5,372 + 4 = 5,376

C.
14. **146 tickets** $584 ÷ $4 = 146
15. **$23** $2,162 ÷ 94 = $23
16. **15 hours**   1,125 ÷ 75 = 15
17. **$372** $8,928 ÷ 24 = $372
18. **160 pages**   960 ÷ 6 = 160

Practice 5 (page 51)

A.
1. 760
2. 3,000
3. 19,600

4. 6,000,000
5. 14,900

B. Your estimate may vary if you rounded to a different place-value column.
6. estimate: 400 + 200 + 200 = **800**   exact: **798**
7. estimate: 5,000 − 1,500 = **3,500**   exact: **3,754**
8. estimate: 800 × 20 = **16,000**   exact: **15,120**
9. estimate: 6,000 ÷ 10 = **600**   exact: **522**

C. Estimates may vary.
10. estimate: $1,000 ÷ 10 = **$100**
    exact: $1,032 ÷ 12 = **$86**
11. estimate: 80,000 + 7,000 = **87,000 square miles**
    exact: 79,617 + 7,326 = **86,943 square miles**
12. estimate: 300 × $20 = **$6,000**
    exact: 280 × $17 = **$4,760**
13. estimate: $28,000 − $18,000 = **$10,000**
    exact: $28,000 − $17,642 = **$10,358**
14. estimate: $60 × 50 = **$3,000**
    exact: $63 × 48 = **$3,024**

Practice 6 (page 53)

A.
1. estimate: $20 + $60 + $10 = **$90**
   exact: **$103**
2. estimate: 40,000 − 10,000 = **30,000**
   exact: **32,564**
3. estimate: 500 × 20 = **10,000**
   exact: **11,960**
4. estimate: 1,000 ÷ 10 = **100**
   exact: **113**

B. Your answer may vary if you chose different compatible numbers.
5. estimate: 180 ÷ 60 = **3**   exact: **3**
6. estimate: 3,600 ÷ 40 = **90**   exact: **84**
7. estimate: 2,000 ÷ 50 = **40**   exact: **42**

C. Your estimate may vary depending on the method you used.
8. estimate: $40 × 10 = **$400**
   exact: $39 × 12 = **$468**
9. estimate: 400 + 300 + 200 = **900 seats**
   exact: 356 + 278 + 186 = **820 seats**
10. estimate: $900,000 − $600,000 = **$300,000**
    exact: $869,430 − $614,500 = **$254,930**
11. estimate: $2,000,000 ÷ 10 = **$200,000**
    exact: $2,025,000 ÷ 9 = **$225,000**
12. estimate: 250 − 150 = **100 tickets**
    exact: 246 − 152 = **94 tickets**

GED Math Practice • Deciding Which Operation to Use (page 55)

1. **(1) Multiply 4 by 40.**   The problem asks for how many books <u>in all</u>. You need to multiply.
2. **(2) Divide $2 by 4.**   The problem gives the total price and asks for the price <u>for each</u> can. You need to divide.
3. **(4) 2 × (62 + 55)**   The problem asks for twice the total of the two days, so you need to add the tips earned each day and multiply the whole amount by 2.
4. **(2) Subtract: 18 minus 12.**   The problem is asking for the difference between the length and width. You need to subtract
5. **(4) 2 × 12 + 2 × 18**   The picture is a rectangle. Therefore, the widths are equal, and the lengths are equal. Multiply the width by 2, and multiply the length by 2. Add to get the total distance around the figure.

GED Math Practice • Calculator Operations and Grid Basics (page 57)

1. **590**
2. **189**
3. **57,400**
4. **18**

5. **124**   Check to see that your calculator uses the correct order of operations. Multiply 18 × 5 first. Then add 34 to the answer.
6. **8**   Subtract 8 from 64 first. Then divide the answer by 7.
7. **(3) 195**   First, multiply the number of minutes by the number of days: 45 × 3 and 30 × 2. Then add the two amounts: 135 + 60 = 195.

Your grid may be slightly different since your answer can start in any column.

8. **$22** First find the cost for each kind of fruit. The cherries cost 2 × $3 = $6, the kiwis cost $1, and the strawberries cost 3 × $5 = $15. Add to get the total cost: $6 + $1 + $15 = $22.

9. **$2** First find the total cost for each kind of fruit. For the cherries, multiply: 4 × $3 = $12. For the strawberries, multiply: 2 × $5 = $10. Subtract the amounts: $12 − $10 = $2.

GED Math Review (page 58)
Part I

1. **(5) 44** Think about one mile of road divided into 120-foot lengths: 5,280 ÷ 120 = 44.

2. **(2) $108** First add the amounts she will spend: $45 + $37 + $60 = $142. Then subtract the answer from the amount in her checking account: $250 − $142 = $108.

3. **(1) $307** First find the total amount for the labor: 3 × $40 = $120. Then add the cost of materials: $120 + $187 = $307.

4. **(3) 35 × 5 + 62 × 3** Multiply the number of tickets by the price of each ticket: 35 × 5 for the adults' tickets, and 62 × 3 for the children's tickets. Add the results.

Your grid may be slightly different since your answer can start in any column.

5. **18** Find the distance from Nogal Canyon through Shirley to Sanchez Pass: 56 + 27 = 83. Then find the distance going through Hart: 43 + 22 = 65. Subtract to find the difference: 83 − 65 = 18 miles.

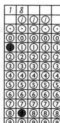

6. **148** Add all four distances: 22 + 27 + 56 + 43 = 148 miles.

Part II

7. **(4) 1,124** Add the number of video rentals for Friday, Saturday, and Sunday: 352 + 566 + 206 = 1,124.

8. **(3) $2,754** Add the number of videos for the two largest rental days: 352 + 566 = 918. Multiply the amount by $3: 918 × $3 = $2,754.

9. **(2) 30 × 2 + 35 + 4** Write the cost of the black ink cartridges: 2 × 30. Then add the cost of the color cartridge, 35, and the shipping, 4.

10. **(1) $150** First divide $21,000 by 12 to get the monthly salary for the new job: $21,000 ÷ 12 = $1,750. Then subtract: $1,750 − $1,600 = $150.

11. **(5) 980** First find the number of gallons used per day: 35 × 4 = 140. Then find the number of gallons used per week: 140 × 7 = 980.

12. **(5) 14** Divide the number of square feet of the roof by 100: 1,200 ÷ 100 = 12. Add the 2 extra packages: 12 + 2 = 14.

PROGRAM 22: DECIMALS
Practice 1 (page 63)

A. 1. a. 9   2. a. 7
    b. 8   b. 6
    c. 3   c. 1

B. 3. four and nine tenths
   4. one hundred twenty-five thousandths
   5. sixteen and eighteen thousandths
   6. thirty-two hundredths
   7. seven and five hundred nine thousandths
   8. fifty and eight thousandths

C. 9. 8.6   12. 1.36
   10. 0.075   13. 6.081
   11. 10.57   14. 11.9

Practice 2 (page 65)

A. 1. 15.7   5. 16.1
   2. $4.13   6. $10
   3. 4   7. 0.51
   4. 0.49   8. $0.98

B. 9. 14.9 < 19.4   14. 0.783 > 0.739
  10. 1.65 > 1.63   15. 0.045 < 0.05
  11. 2.80 = 2.8   16. 8.315 > 7.915
  12. 0.425 > 0.325   17. 4.60 = 4.6
  13. 3.56 < 3.561   18. 10.032 < 10.302

C. 19. **$46.60, $46.50, $42.60** $46 is greater than $42, so $42.60 is the smallest amount. $0.60 is greater than $0.50, so $46.60 is greater than $46.50.

  20. **5.8 grams** Since the digit to the right of the tenths place is less than five, round down.

  21. **Armando, Ken, Roy, James** Work from left to right. Remember, the greatest time is the slowest, or fourth-place, time.

  22. **Books A, C, and D** The call numbers must be equal to or between 731.18 and 805.95. Both 750.04 and 802.96 are greater than 731.18 and less than 805.95. Book D has a call number equal to the lowest number in the range.

  23. **59 by 23 meters** 59.47 rounds down to 59, and 22.87 rounds up to 23

Practice 3 (page 67)

A. 1. 9.3   4. $36.51
   2. 5.5   5. 22.2
   3. 8.19   6. 7.14

B. 7. 8.17   10. 13.416
   8. $2.34   11. 5.16
   9. 11.397   12. 167.05

C. 13. **16.205 inches** $6.48 + 5.6 + 4.125 = 16.205$
   14. **3.3 centimeters** $0.4 + 2.5 + 0.4 = 3.3$
   15. **$8.22** $6.95 + $1.27 = 8.22
   16. **8.1 miles** $2.7 + 3.6 + 1.8 = 8.1$
   17. **53.375 meters** $24.5 + 13.25 + 15.625 = 53.375$
   18. **9.79 pounds** $5.44 + 2.5 + 1.85 = 9.79$

Practice 4 (page 69)

A. 1. **0.46** Check: $0.46 + 0.48 = 0.94$
   2. **2.8** Check: $2.8 + 1.92 = 4.72$
   3. **10.421** Check: $10.421 + 4.705 = 15.126$
   4. **18.85** Check: $18.85 + 6.85 = 25.7$
   5. **0.63** Check: $0.63 + 6.9 = 7.53$
   6. **14.915** Check: $14.915 + 3.125 = 18.04$
   7. **36.25** Check: $36.25 + 13.75 = 50$
   8. **12.02** Check: $12.02 + 0.48 = 12.5$
   9. **20.108** Check: $20.108 + 15.9 = 36.008$
B. 10. **7.82** Check: $7.82 + 3.68 = 11.5$
   11. **$12.66** Check: $12.66 + $5.34 = 18
   12. **70.755** Check: $70.755 + 4.25 = 75.005$
   13. **$69.37** Check: $69.37 + $34.50 = 103.87
   14. **1.491** Check: $1.491 + 0.009 = 1.5$
   15. **5.969** Check: $5.969 + 12.4 = 18.369$
C. 16. **3,830.4 miles** Subtract: $16,309.2 - 12,478.8 = 3,830.4$ miles.
   17. **0.475 inch** Subtract the smaller amount from the larger amount: $1.275 - 0.8 = 0.475$.
   18. **$12.08** Subtract the cost of the purchase from the amount Max paid. Two $20 bills equal $40, and $40 - $27.92 = 12.08.
   19. **$889.59** First, add the two check amounts: $538.70 + $45 = 583.70. Subtract the check total from the amount in Jesse's checking account. $1,346.87 - $583.70 = 763.17. Finally, add the amount Jesse deposited: $763.17 + $126.42 = 889.59.
   20. **4.6 degrees** Subtract the normal body temperature from Stuart's current temperature: $103.2 - 98.6 = 4.6$.
   21. **0.3 point** First, add the points CFH gained each day to find the total gain. $1.4 + 4.8 + 2.5 = 8.7$. Then, subtract the total amount CFH gained from the total amount KRO gained. $9 - 8.7 = 0.3$.

Practice 5 (page 71)

A. 1. 14
   2. 1.46
   3. 0.04
   4. $8.40
   5. 352.8
   6. **$6.07** The product 6.072 rounds to $6.07.
   7. 0.503
   8. 0.00574
   9. **$2.01** The product 2.0083 rounds to $2.01.

B. 10. 2.892   13. 0.623
   11. 0.0345   14. 100.15
   12. 3.186   15. 1.204

C. 16. **$2,167.80** Multiply $180.65 by 12.
   17. **68.4 pounds** Multiply 7.6 by 9.
   18. **$28.88** Multiply 82.5 by $0.35. Round the product 28.875 to $28.88.
   19. **$15.75** Multiply $900 by 0.0175.
   20. **32.5 pounds** Multiply 8.125 by 4.
   21. **33.6 pounds** Multiply 4.2 by 8.
   22. **20.52 pounds** Add the weight of each order: $2.64 + 4.2 = 6.84$ pounds. Multiply by 3: $6.84 \times 3 = 20.52$ pounds.

Practice 6 (page 73)

A. 1. 3.6   6. 223
   2. 1.25   7. 605
   3. 3.18   8. 480
   4. 2.16   9. 26
   5. 0.09

B. 10. **$2.33** 2.333 rounds to $2.33
   11. **$0.79** 0.785 rounds to $0.79
   12. **$76.92** 76.923 rounds to $76.92
   13. **$0.58** 0.582 rounds to $0.58
C. 14. **$12.56** Divide his pay by the number of hours he worked: $113.04 \div 9 = 12.56.
   15. **16 bottles** Divide 24 liters, the amount in the barrel, by 1.5 liters, the amount that one bottle will hold.
   16. **12.5 miles** Divide her total mileage by the number of days: $74.7 \div 6 = 12.45$, which rounds to 12.5 miles.
   17. **$7.95** Divide the price of the roast by the number of pounds: $41.34 \div 5.2 = 7.95.
   18. **15 plants** Divide $20 by $1.29, the price of a 3-inch plant. Since you can't buy part of a plant, stop dividing once you know the whole-number part of the answer.
   19. **$2.50** Divide $500 by 200. Express the quotient 2.5 as dollars and cents: $2.50.

GED Math Practice • Does the Answer Make Sense? (page 75)

1. **(1) $17.33** The cost per pound is nearly $5. Since $3 \times $5 = 15, and $4 \times $5 = 20, the cost must be between $15 and $20. ($4.95 \times 3.5 = $17.33)
2. **(2) $9.60** The bill should be split 4 ways: Jason plus his 3 friends. Since the total bill is less than $40, each person will pay less than $40 \div 4 = 10. ($38.40 \div 4 = $9.60)
3. **(1) 8.3** Subtract to find the difference. The result would be less than the two numbers being compared. ($26.8 - 18.5 = 8.3$)
4. **(2) 69.6** Round 5.8 ounces to 6 ounces. Multiply the number of ounces by the number of jars: $6 \times 12 = 72$. Since 5.8 is close to 6, the answer should be close to 72. ($5.8 \times 12 = 69.6$)
5. **(5) 11.3** Since you need to subtract, the answer must be small. You can eliminate options (1), (2), and (3) because these numbers are larger than or equal to those in the problem. ($47.2 - 35.9 = 11.3$)

6. (4) 41.8 Round each of the measurements to 10 and multiply: $4 \times 10 = 40$. You can eliminate options (1) and (5). (Add to find the distance: $9.9 + 11.2 + 8.7 + 12 = 41.8$ cm.)

GED Math Practice • Decimals on the Calculator and Grid (page 77)

1. 7.64
2. $133.33 (133.333 rounds to $133.33)
3. 76.95
4. 8.3
5. 17.54
6. 0.8
7. **98.79** Add the cabin rate and the taxes to get the cost for one night: $89.00 + $6.23 + $3.56 = $98.79.

8. **2.20** Use the parentheses keys to find the difference between both sums. $(6.93 + 3.96) - (5.53 + 3.16) = 2.2 = 2.20

GED Math Review (pages 78–79)

Part I

1. **(2) $0.50** Divide the cost of the case by the number of bottles of water: $11.95 ÷ 24 = 0.4979166$. Round the answer to the nearest cent to get $0.50.
2. **(3) $20.75** First find the cost of the two tickets and two drinks: $2 \times $6.50 = 13, and $2 \times $2.00 = 4. Add the three costs: $13 + $4 + $3.75 = $20.75.
3. **(4) $1.50** First find the regular cost of the ticket, medium popcorn, and drink: $6.50 + $3.00 + $2.00 = 11.50. Subtract the discount price from the normal price: $11.50 − $10.00 = $1.50.
4. **(1) $235.63** Multiply Rebecca's hourly rate by the number of hours per day by the number of days: $7.25 \times 6.5 \times 5 = 235.625, which rounds to $235.63.
5. **(5) 32.5** Multiply the number of pounds Joe lost each week by the number of weeks: $2.5 \times 13 = 32.5$ pounds.
6. **(2) 157.5** Multiply the number of inches by 30 miles per inch: $5.25 \times 30 = 157.5$.
7. **1.11** First multiply the cost per pound by the number of pounds of potatoes: $0.39 \times 10 = $3.90. Then subtract the cost of the 10-pound bag: $3.90 − $2.79 = $1.11.

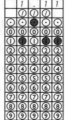

Part II

8. **(4) 4.99, 5.15, 5.20, 5.53, 5.71** Compare the whole numbers first. Then compare the tenths place. Order the times from fastest to slowest. (Note that the fastest time will be the smallest number.)
9. **(3) 0.05** Gloria's best time this week was 4.99 minutes. Subtract the two times: $5.04 − 4.99 = 0.05$.
10. **(5) $3.5 \times 60 + 1.5 \times 30$** Multiply each time and distance: 3.5×60 and 1.5×30 for the construction zone miles. Add the two results.
11. **(3) $0.015** Divide the cost of the paper by the number of sheets of paper: $7.50 ÷ 500 = $0.015.
12. **(3) $1,250** First multiply to find Jim's parents' contribution: $500 \times $1.50 = 750. Then add the result to Jim's savings: $750 + $500 = $1,250.
13. **(4) $78.75** Multiply the number of quarters by $0.25: $315 \times $0.25 = $78.75.
14. **437.5** Multiply the number of miles per gallon by the number of gallons in the gas tank: $25 \times 17.5 = 437.5$ miles.

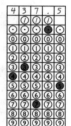

PROGRAM 23: FRACTIONS

Practice 1 (page 83)

A. 1. $\frac{1}{2}$ 3. $\frac{5}{8}$

    2. $\frac{3}{4}$

B. 4. $\frac{1}{4}$ 5. $\frac{2}{3}$

C. 6. $\frac{1}{2}$ Divide both parts of the fraction by 3.

   7. $\frac{3}{5}$ Divide both parts of the fraction by 4.

   8. $\frac{1}{3}$ Divide both parts of the fraction by 8.

   9. $\frac{3}{4}$ Divide both parts of the fraction by 10.

  10. $\frac{2}{9}$ Divide both parts of the fraction by 3.

  11. $\frac{2}{5}$ Divide both parts of the fraction by 5.

D. 12. $\frac{3}{5}$ Divide both parts of the fraction by 25. $\frac{75}{125}$ simplifies to $\frac{3}{5}$

  13. $\frac{1}{4}$ Divide both parts of the fraction by $30. $\frac{$30}{120}$ simplifies to $\frac{1}{4}$

Practice 2 (page 85)

A. 1. $1\frac{5}{10} = 1\frac{1}{2}$ 5. $2\frac{3}{6} = 2\frac{1}{2}$

   2. $2\frac{2}{9}$ 6. 3

   3. 2   7. $8\frac{4}{4} = 8\frac{1}{2}$

   4. $10\frac{1}{2}$ 8. $4\frac{9}{12} = 4\frac{3}{4}$

B. 9. $\frac{5}{8}$ 13. $\frac{10}{6} = 1\frac{4}{6} = 1\frac{2}{3}$

  10. $\frac{5}{4} = 1\frac{1}{4}$ 14. $\frac{10}{5} = 2$

  11. $\frac{4}{11}$ 15. $\frac{6}{20} = \frac{3}{10}$

  12. $\frac{5}{15} = \frac{1}{3}$ 16. $\frac{4}{14} = \frac{2}{7}$

C. 17. $1\frac{1}{5}$ gallons

$\frac{3}{10} + \frac{4}{10} + \frac{5}{10} = \frac{12}{10} = 1\frac{2}{10} = 1\frac{1}{5}$

18. $\frac{1}{2}$ mile

$\frac{7}{8} - \frac{3}{8} = \frac{4}{8} = \frac{1}{2}$

19. $1\frac{1}{6}$ pounds

$\frac{5}{12} + \frac{9}{12} = \frac{14}{12} = 1\frac{2}{12} = 1\frac{1}{6}$

20. $\frac{3}{8}$ acre

$\frac{15}{16} - \frac{9}{16} = \frac{6}{16} = \frac{3}{8}$

Practice 3 (page 87)

A. 1. $\frac{1}{6} + \frac{2}{3} = \frac{1}{6} + \frac{4}{6} = \frac{5}{6}$

2. $\frac{2}{7} + \frac{3}{14} = \frac{4}{14} + \frac{3}{14} = \frac{7}{14} = \frac{1}{2}$

3. $\frac{1}{8} + \frac{5}{6} = \frac{3}{24} + \frac{20}{24} = \frac{23}{24}$

4. $\frac{2}{3} + \frac{8}{15} = \frac{10}{15} + \frac{8}{15} = \frac{18}{15} = 1\frac{3}{15} = 1\frac{1}{5}$

5. $\frac{2}{3} + \frac{3}{4} = \frac{8}{12} + \frac{9}{12} = \frac{17}{12} = 1\frac{5}{12}$

6. $\frac{4}{5} + \frac{2}{7} = \frac{28}{35} + \frac{10}{35} = \frac{38}{35} = 1\frac{3}{35}$

7. $\frac{1}{3} + \frac{1}{4} + \frac{1}{2} = \frac{4}{12} + \frac{3}{12} + \frac{6}{12} = \frac{13}{12} = 1\frac{1}{12}$

8. $\frac{2}{5} + \frac{7}{20} + \frac{9}{10} = \frac{8}{20} + \frac{7}{20} + \frac{18}{20} = \frac{33}{20} = 1\frac{13}{20}$

9. $\frac{11}{12} - \frac{1}{2} = \frac{11}{12} - \frac{6}{12} = \frac{5}{12}$

10. $\frac{7}{15} - \frac{2}{5} = \frac{7}{15} - \frac{6}{15} = \frac{1}{15}$

11. $\frac{17}{20} - \frac{1}{4} = \frac{17}{20} - \frac{5}{20} = \frac{12}{20} = \frac{3}{5}$

12. $\frac{5}{8} - \frac{5}{16} = \frac{10}{16} - \frac{5}{16} = \frac{5}{16}$

13. $\frac{7}{12} - \frac{3}{8} = \frac{14}{24} - \frac{9}{24} = \frac{5}{24}$

14. $\frac{5}{9} - \frac{1}{4} = \frac{20}{36} - \frac{9}{36} = \frac{11}{36}$

15. $\frac{9}{10} - \frac{3}{15} = \frac{27}{30} - \frac{6}{30} = \frac{21}{30} = \frac{7}{10}$

B. 16. $\frac{5}{16}$ inch   Subtract to find the difference.

$\frac{13}{16} - \frac{1}{2} = \frac{13}{16} - \frac{8}{16} = \frac{5}{16}$ inch

17. $\frac{5}{24}$ yard   Subtract to find how much ribbon
is left. $\frac{5}{6} - \frac{5}{8} = \frac{20}{24} - \frac{15}{24} = \frac{5}{24}$ yard

18. $1\frac{7}{12}$ cups   Add to find the total amount.

$\frac{3}{4} + \frac{1}{3} + \frac{1}{2} = \frac{9}{12} + \frac{4}{12} + \frac{6}{12} = \frac{19}{12} = 1\frac{7}{12}$ cups

Fractions • Practice 4 (page 89)

A. 1. $5\frac{3}{3} = 6$

2. $8\frac{8}{12} = 8\frac{2}{3}$

3. $12\frac{9}{6} = 13\frac{3}{6} = 13\frac{1}{2}$

4. $15\frac{15}{10} = 16\frac{5}{10} = 16\frac{1}{2}$

5. $10\frac{17}{18}$

6. $5\frac{37}{30} = 6\frac{7}{30}$

7. $6\frac{21}{20} = 7\frac{1}{20}$

8. $13\frac{9}{12} = 13\frac{3}{4}$

9. $5\frac{4}{8} = 5\frac{1}{2}$

10. $4\frac{3}{10}$

11. $7\frac{5}{7}$

12. $1\frac{3}{10}$

13. $1\frac{13}{24}$

14. $\frac{5}{6}$

15. $4\frac{19}{21}$

16. $6\frac{5}{6}$

B. 17. $1\frac{3}{20}$ points   Subtract:

$8\frac{9}{10} - 7\frac{3}{4} = 8\frac{18}{20} - 7\frac{15}{20} = 1\frac{3}{20}$

18. $16\frac{1}{4}$ feet   Add the three measurements:

$6\frac{3}{4} + 6\frac{3}{4} + 2\frac{3}{4} = 14\frac{9}{4} = 14 + 2\frac{1}{4} = 16\frac{1}{4}$

19. $7\frac{1}{3}$ feet   Subtract: $9 - 1\frac{2}{3} = 8\frac{3}{3} - 1\frac{2}{3} = 7\frac{1}{3}$

20. $19\frac{7}{8}$ feet   Add to find how much you will use:

$5\frac{5}{8} + 4\frac{1}{2} = 5\frac{5}{8} + 4\frac{4}{8} = 9\frac{9}{8} = 9 + 1\frac{1}{8} = 10\frac{1}{8}$ feet.

Subtract from the length of the roll:

$30 - 10\frac{1}{8} = 29\frac{8}{8} - 10\frac{1}{8} = 19\frac{7}{8}$ feet.

21. $9\frac{3}{4}$ hours   Find the total hours he has worked:

$2\frac{1}{2} + 4 + 3\frac{3}{4} = 2\frac{2}{4} + 4 + 3\frac{3}{4} = 9\frac{5}{4} = 10\frac{1}{4}$

Subtract: $20 - 10\frac{1}{4} = 19\frac{4}{4} - 10\frac{1}{4} = 9\frac{3}{4}$ hours

Practice 5 (page 91)

A. 1. $\frac{7}{16}$

2. $\frac{18}{50} = \frac{9}{25}$

3. $\frac{24}{36} = \frac{2}{3}$

4. $\frac{3}{30} = \frac{1}{10}$

5. $\frac{1}{16}$

6. $\frac{9}{2} = 4\frac{1}{2}$

7. 80

8. 2

9. 10

10. $\frac{49}{4} = 12\frac{1}{4}$

11. $\frac{52}{10} = 5\frac{2}{10} = 5\frac{1}{5}$ or $\frac{26}{5} = 5\frac{1}{5}$

12. $\frac{93}{6} = 15\frac{3}{6} = 15\frac{1}{2}$ or $\frac{31}{2} = 15\frac{1}{2}$

13. 5

14. $\frac{45}{4} = 11\frac{1}{4}$

15. 10

B. 16. $\frac{1}{10}$ Multiply: $\frac{\overset{1}{\cancel{2}}}{5} \times \frac{1}{\underset{2}{\cancel{4}}} = \frac{1}{10}$

17. $\$84$ $\$112 \times \frac{3}{4} = \frac{\overset{28}{\cancel{112}}}{1} \times \frac{3}{\underset{1}{\cancel{4}}} = \frac{84}{1} = \$84$

18. 175 miles $70 \times 2\frac{1}{2} = \frac{\overset{35}{\cancel{70}}}{1} \times \frac{5}{\underset{1}{\cancel{2}}} = \frac{175}{1} = 175$

19. $27\frac{1}{2}$ miles   Add to find the length of the route:

$1\frac{3}{4} + 1\frac{1}{2} + 2\frac{1}{4} = 5\frac{1}{2}$

Multiply by 5: $5\frac{1}{2} \times 5 = \frac{11}{2} \times \frac{5}{1} = \frac{55}{2} = 27\frac{1}{2}$

20. $7\frac{1}{5}$ pounds $1\frac{4}{5} \times 4 = \frac{9}{5} \times \frac{4}{1} = \frac{36}{5} = 7\frac{1}{5}$

Practice 6 (page 93)

A. 1. $\frac{3}{4}$ 9. $\frac{2}{7}$

2. $2\frac{1}{2}$ 10. $2\frac{1}{2}$

3. $\frac{2}{3}$ 11. $1\frac{1}{2}$

4. $5\frac{1}{2}$ 12. 3

5. $\frac{2}{3}$ 13. $\frac{2}{3}$

6. 25   14. $5\frac{3}{4}$

7. 18   15. $2\frac{7}{9}$

8. $\frac{1}{5}$

B. 16. **16 dresses**

$34 \div 2\frac{1}{8} = \frac{34}{1} \div \frac{17}{8} = \frac{34}{1} \times \frac{8}{17} = \frac{\overset{2}{34}}{1} \times \frac{8}{\overset{}{17}} = \frac{16}{1} = 16$

17. **56 lots**

$21 \div \frac{3}{8} = \frac{21}{1} \div \frac{3}{8} = \frac{21}{1} \times \frac{8}{3} = \frac{\overset{7}{21}}{1} \times \frac{8}{\overset{}{3}} = \frac{56}{1} = 56$

18. **50 miles per hour**

$175 \div 3\frac{1}{2} = \frac{175}{1} \div \frac{7}{2} = \frac{175}{1} \times \frac{2}{7} = \frac{\overset{25}{175}}{1} \times \frac{2}{\overset{}{7}} = \frac{50}{1} = 50$

19. **28 books**

$35 \div 1\frac{1}{4} = \frac{35}{1} \div \frac{5}{4} = \frac{35}{1} \times \frac{4}{5} = \frac{\overset{7}{35}}{1} \times \frac{4}{\overset{}{5}} = \frac{28}{1} = 28$

20. **46 one-eighth cups**

$5\frac{3}{4} \div \frac{1}{8} = \frac{23}{4} \div \frac{1}{8} = \frac{23}{4} \times \frac{8}{1} = \frac{23}{\overset{}{4}} \times \frac{\overset{2}{8}}{1} = \frac{46}{1} = 46$

21. **8 troubleshooting calls**

$6 \div \frac{3}{4} = \frac{6}{1} \div \frac{3}{4} = \frac{6}{1} \times \frac{4}{3} = \frac{\overset{2}{6}}{1} \times \frac{4}{\overset{}{3}} = \frac{8}{1} = 8$

22. **20 requests for refunds**

$6\frac{2}{3} \div \frac{1}{3} = \frac{20}{3} \div \frac{1}{3} = \frac{20}{3} \times \frac{3}{1} = \frac{20}{\overset{}{3}} \times \frac{\overset{1}{3}}{1} = \frac{20}{1} = 20$

GED Math Practice • The Information You Need (page 95)

1. **compute needed information** Add the bills for electricity and water: $35.95 + $22.40 = $58.35. Then subtract from the amount for rent: $450.00 − $58.35 = $391.65.

2. **not enough** You need to know how many people went to the reception.

3. **compute needed information** Add the number of Democrats and the number of Republicans to get the denominator: 1,500 + 2,000 = 3,500. The numerator is the number of Democrats (2,000). Write the fraction, and reduce to lowest terms: $\frac{2,000}{3,500} = \frac{20}{35} = \frac{4}{7}$

4. **too much** Add the two lengths and two widths. You do not need the diagonal measurement.

5. **not enough** You need to know how much space is needed for one rosebush.

GED Math Practice • Fractions on the Calculator and Grid (page 97)

1. 8   4. $\frac{1}{16}$

2. 3   5. $\frac{13}{16}$

3. $4\frac{1}{3}$ 6. $4\frac{7}{10}$

7. $\frac{5}{12}$ First find the common denominator for the fractions for rent and utilities. Then add: $\frac{1}{4} + \frac{1}{6} = \frac{3}{12} + \frac{2}{12} = \frac{5}{12}$.

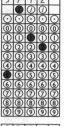

8. **375** Multiply the total budget by the fraction for the car payment: $1,500 × $\frac{1}{4}$ = $375.

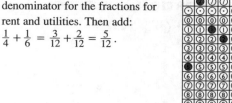

GED Math Review (pages 98–99)
Part I

1. **(5) $8\frac{1}{4}$** Multiply the number of yards by the number of feet per yard: $2\frac{3}{4} \times 3 = 8\frac{1}{4}$.

2. **(1) 5** Find the difference between the length and the width by subtracting: $16\frac{1}{2} - 11\frac{1}{2} = 5$.

3. **(5) Not enough information is given.** To find the distance Kaneesha is from the start, you need to know the entire distance around the trail.

4. **(4) $\frac{7}{8}$** Subtract the smaller number from the larger number: $24\frac{1}{16} - 23\frac{3}{16} = \frac{14}{16} = \frac{7}{8}$.

5. **(3) $\frac{1}{2}$** Multiply the number of cups of milk in the recipe by $\frac{1}{3}$: $1\frac{1}{2} \times \frac{1}{3} = \frac{3}{2} \times \frac{1}{3} = \frac{3}{6} = \frac{1}{2}$.

6. **8** To find the number of logs, divide the height of the wall by the diameter of the logs: $58 \div 7\frac{1}{4} = 58 \div \frac{29}{4} = \frac{\overset{2}{58}}{1} \times \frac{4}{\overset{}{29}} = \frac{8}{1} = 8$.

Part II

7. **(2) $\frac{1}{4}$** The numerator is the number of hours Monica worked. The denominator is the total number of hours. The fraction is $\frac{35}{140} = \frac{1}{4}$.

8. **(4) $13\frac{3}{4}$** Subtract Tom's hours from Frank's hours: $40 - 26\frac{1}{4} = 39\frac{4}{4} - 26\frac{1}{4} = 13\frac{3}{4}$.

9. **(2) $5\frac{1}{4}$** To triple the recipe, multiply the cups of flour by 3: $1\frac{3}{4} \times 3 = 3\frac{9}{4} = 5\frac{1}{4}$.

10. **(3) $(\frac{1}{2} + \frac{3}{4}) \times 6$** Find the amount of fabric he needs for each chair seat and back: $\frac{1}{2} + \frac{3}{4}$. Multiply this entire amount for 6 chairs: $(\frac{1}{2} + \frac{3}{4}) \times 6$.

11. **(4) $6.25** Multiply the amount Nathan started with by $\frac{1}{2}$. $50 × $\frac{1}{2}$ = $25. He has $25 left after buying groceries. He spent $\frac{3}{4}$ of this amount, so will have $\frac{1}{4}$ left after Friday's dinner. Multiply: $25 × $\frac{1}{4} = \frac{$25}{4} = 6\frac{1}{4}$, or $6.25.

12. **30** Divide the total weight of ground beef by the weight of a burger: $7\frac{1}{2} \div \frac{1}{4} = \frac{15}{\overset{}{2}} \times \frac{\overset{2}{4}}{1} = \frac{30}{1} = 30$.

PROGRAM 24: RATIO, PROPORTION, AND PERCENT

Practice 1 (page 103)

A. 1. $\frac{21}{3} = \frac{7}{1}$ 6. $\frac{100}{25} = \frac{4}{1}$

2. $\frac{12}{15} = \frac{4}{5}$ 7. $\frac{42}{30} = \frac{7}{5}$

3. $\frac{4}{7}$ 8. $\frac{25}{40} = \frac{5}{8}$

4. $\frac{6}{27} = \frac{2}{9}$ 9. $\frac{11}{121} = \frac{1}{11}$

5. $\frac{16}{9}$

B. 10. $\frac{\$240}{\$180} = \frac{4}{3}$

11. $\frac{4,500}{1,800} = \frac{5}{2}$

12. $\frac{5}{8}$ Add to find the total time: $25 + 15 = 40$ hours. Write the ratio and simplify: $\frac{25}{40} = \frac{5}{8}$

13. $\frac{5}{4}$ Subtract to find the take-home pay: $\$270 - \$54 = \$216$. Write the ratio and simplify: $\frac{\$270}{\$216} = \frac{5}{4}$

14. $\frac{21}{18} = \frac{7}{6}$

15. $\frac{1}{2}$ Add to find the total games played: 15 wins + 15 losses = 30 games. Write the ratio and simplify: $\frac{15}{30} = \frac{1}{2}$

16. **Kryptonite**   Only the Kryptonite and Bandits have more wins than losses, so these are the only possible choices. Check each.

Kryptonite: $\frac{21}{9} = \frac{7}{3}$ Bandits: $\frac{18}{12} = \frac{3}{2}$

17. $\frac{24}{12} = \frac{2}{1}$

18. **High Heat**   Only the High Heat team has more losses than wins. It is the only possible answer. $\frac{24}{6} = \frac{4}{1}$

Practice 2 (page 105)

A. 1. 180 calories per serving
2. 6 customers per hour
3. $12 per square yard of carpet
4. 440 feet per minute
5. $\frac{1}{2}$ ounce (or 0.5 ounce) per bag
6. $15 per hour
7. 12 cents per ounce
8. 14 children per team
9. 150 meters per minute
10. 55 miles per hour

B. 11. **Ache-Away**   Find the unit rates. Flu Tabs: $\$6.96 \div 60 = \$0.116$, or 11.6 cents per tablet. Ache-Away: $\$4.16 \div 40 = \$0.104$, or 10.4 cents per tablet. Ache-Away is less expensive.

12. **BestInk**   Find the unit rates.
FineLine: $\$7.12 \div 8 = \$0.89$ per pen
BestInk: $\$10.32 \div 12 = \$0.86$ per pen
WriteAway: $\$9.20 \div 10 = \$0.92$ per pen
BestInk is least expensive per pen.

C. 13. $\$3.90 \div 5 = $ **$0.78 per pound**
14. $\$7,500 \div 6 = $ **$1,250 per year**
15. $\$1.12 \div 16 = $ **$0.07 per minute**

16. $275 \div 5 = $ **55 miles per hour**
17. $360 \div 15 = $ **24 miles per gallon**
18. **3-ream package**   Find the unit rate per ream for each package.
3-ream package: $\$21.60 \div 3 = \$7.20$
5-ream package: $\$36.25 \div 5 = \$7.25$
19. **$0.04 or 4 cents**   Find the unit rate for each pack.
5-pack: $\$3.90 \div 5 = \$0.78$ per battery
8-pack: $\$5.92 \div 8 = \$0.74$ per battery
The problem asks how much less Min Lee will pay per battery. Subtract to find the difference: $\$0.78 - \$0.74 = \$0.04$, or 4 cents.

Practice 3 (page 107)

A. 1. $3 \times 10 = 30$; $30 \div 2 = $ **15**
2. $3 \times 20 = 60$; $60 \div 5 = $ **12**
3. $8 \times 35 = 280$; $280 \div 7 = $ **40**
4. $6 \times 9 = 54$; $54 \div 4 = $ **13.5** or $13\frac{1}{2}$
5. $11 \times 27 = 297$; $297 \div 9 = $ **33**
6. $45 \times 10 = 450$; $450 \div 25 = $ **18**
7. $24 \times 6 = 144$; $144 \div 4 = $ **36**
8. $3 \times 28 = 84$; $84 \div 4 = $ **21**
9. $12 \times 13 = 156$; $156 \div 26 = $ **6**
10. $1.5 \times 16 = 24$; $24 \div 6 = $ **4**
11. $0.5 \times 32 = 16$; $16 \div 8 = $ **2**
12. $5 \times 2 = 10$: $10 \div 25 = $ **0.4** or $\frac{2}{5}$

B. 13. **$264** $\frac{4}{\$33} = \frac{32}{x}$
$33 \times 32 = \$1,056$; $\$1,056 \div 4 = \$264$

14. **300 products** $\frac{25}{5} = \frac{x}{60}$
$25 \times 60 = 1,500$; $1,500 \div 5 = 300$

15. **390 miles** $\frac{156}{6} = \frac{x}{15}$
$156 \times 15 = 2,340$; $2,340 \div 6 = 390$

16. **3 scoops** $\frac{2}{6} = \frac{x}{9}$
$2 \times 9 = 18$; $18 \div 6 = 3$

Practice 4 (page 109)

1. **$85.00** $\frac{3 \text{ titles}}{\$25.50} = \frac{10 \text{ titles}}{x}$
$\$25.50 \times 10 = \$255$; $\$255 \div 3 = \$85.00$

2. **4 ounces** $\frac{1 \text{ part cleaner}}{8 \text{ parts water}} = \frac{x \text{ ounces cleaner}}{32 \text{ ounces water}}$
$1 \times 32 = 32$; $32 \div 8 = 4$ ounces

3. **$8\frac{3}{4}$ quarts** $\frac{1 \text{ part cleaner}}{5 \text{ parts water}} = \frac{1\frac{1}{4} \text{ quarts cleaner}}{x \text{ quarts water}}$
$5 \times 1\frac{3}{4} = 8\frac{3}{4}$; $8\frac{3}{4} \div 1 = 8\frac{3}{4}$ quarts

4. **900 miles** $\frac{\frac{1}{2} \text{ inch}}{200 \text{ miles}} = \frac{2\frac{1}{4} \text{ inches}}{x \text{ miles}}$
$200 \times 2\frac{1}{4} = 450$; $450 \div \frac{1}{2} = 900$ miles

5. **$25.20** $\frac{5 \text{ rolls}}{\$7} = \frac{18 \text{ rolls}}{x}$
$\$7 \times 18 = \$126$; $\$126 \div 5 = \$25.20$

6. **$1\frac{3}{4}$ inches** or **1.75 inches** $\frac{1 \text{ in.}}{80 \text{ mi.}} = \frac{x \text{ in.}}{140 \text{ mi.}}$
$1 \times 140 = 140$; $140 \div 80 = 1.75$ or $1\frac{3}{4}$ inches

7. **21 ounces of white**
$\frac{7 \text{ parts white}}{2 \text{ parts green oxide}} = \frac{x \text{ ounces white}}{6 \text{ ounces green oxide}}$
$7 \times 6 = 42$; $42 \div 2 = 21$ ounces of white

8. **10 jerseys** $\frac{4 \text{ jerseys}}{\$50} = \frac{x \text{ jerseys}}{\$125}$
$4 \times \$125 = \$500$; $\$500 \div \$50 = 10$ jerseys

Practice 5 (page 111)

A. **1.** **63** $\frac{x}{90} = \frac{70}{100}$
$90 \times 70 = 6,300$; $6,300 \div 100 = 63$

2. **12%** $\frac{\$60}{\$500} = \frac{x}{100}$
$\$60 \times 100 = \$6,000$; $\$6,000 \div \$500 = 12$;
$\frac{12}{100} = 12\%$

3. **20** $\frac{3}{x} = \frac{15}{100}$
$3 \times 100 = 300$; $300 \div 15 = 20$

4. **\$6** $\frac{x}{\$8} = \frac{75}{100}$
$\$8 \times 75 = \$600$; $\$600 \div 100 = \$6$

5. **25%** $\frac{\$9}{\$36} = \frac{x}{100}$
$\$9 \times 100 = \$900$; $\$900 \div \$36 = 25$; $\frac{25}{100} = 25\%$

6. **24** $\frac{x}{80} = \frac{30}{100}$
$80 \times 30 = 2,400$; $2,400 \div 100 = 24$

7. **38** $\frac{19}{x} = \frac{50}{100}$
$19 \times 100 = 1,900$; $1,900 \div 50 = 38$

8. **88%** $\frac{22}{25} = \frac{x}{100}$
$22 \times 100 = 2,200$; $2,200 \div 25 = 88$; $\frac{88}{100} = 88\%$

B. **9.** **90%** $\frac{18}{20} = \frac{x}{100}$
$18 \times 100 = 1,800$; $1,800 \div 20 = 90$; $\frac{90}{100} = 90\%$

10. **26 cars** $\frac{x}{40} = \frac{65}{100}$
$40 \times 65 = 2,600$; $2,600 \div 100 = 26$ cars

11. **880 people were surveyed** $\frac{132}{x} = \frac{15}{100}$
$132 \times 100 = 13,200$; $13,200 \div 15 = 880$ people

12. **3%** $\frac{21}{700} = \frac{x}{100}$
$21 \times 100 = 2,100$; $2,100 \div 700 = 3$; $\frac{3}{100} = 3\%$

13. **75%** Add the nuts: $6 + 3 + 2 + 1 = 12$ ounces.
The peanuts and cashews total $6 + 3 = 9$
ounces. Find what percent 9 is of 12: $\frac{9}{12} = \frac{x}{100}$
$9 \times 100 = 900$; $900 \div 12 = 75$; $\frac{75}{100} = 75\%$

Practice 6 (page 113)

1. **\$28** $\frac{x}{\$35} = \frac{20}{100}$
$\$35 \times 20 = \$700$; $\$700 \div 100 = \$7$
Subtract: $\$35 - \$7 = \$28$

2. **25%** Subtract: $\$15,000 - \$12,000 = \$3,000$
$\frac{\text{amount of change}}{\text{original amount}} = \frac{\text{rate}}{100}$ $\frac{\$3,000}{\$12,000} = \frac{x}{100}$
$\$3,000 \times 100 = \$300,000$;
$\$300,000 \div 12,000 = 25$; $\frac{25}{100} = 25\%$

3. **\$72** Find the first sale price: $\frac{x}{\$120} = \frac{25}{100}$
$\$120 \times 25 = \$3,000$; $\$3,000 \div 100 = \$30$
Subtract: $\$120 - \$30 = \$90$
Find the second sale price: $\frac{x}{\$90} = \frac{20}{100}$
$\$90 \times 20 = \$1,800$; $\$1,800 \div 100 = \$18$
Subtract: $\$90 - \$18 = \$72$

4. **20%** Subtract: $150 - 120 = 30$
$\frac{\text{amount of change}}{\text{original amount}} = \frac{\text{rate}}{100}$ $\frac{30}{150} = \frac{x}{100}$
$30 \times 100 = 3,000$; $3,000 \div 150 = 20$; $\frac{20}{100} = 20\%$

5. **\$513.60** $\frac{x}{\$480} = \frac{7}{100}$
$\$480 \times 7 = \$3,360$; $\$3,360 \div 100 = \$33.60$
Add: $\$480 + \$33.60 = \$513.60$

6. **3,289** Find the first increase: $\frac{x}{2,600} = \frac{15}{100}$
$2,600 \times 15 = 39,000$; $39,000 \div 100 = 390$
Add: $2,600 + 390 = 2,990$
Find the second increase: $\frac{x}{2,990} = \frac{10}{100}$
$2,990 \times 10 = 29,900$; $29,900 \div 100 = 299$
Add: $2,990 + 299 = 3,289$

7. **\$1,048.60** $\frac{x}{\$980} = \frac{7}{100}$
$\$980 \times 7 = \$6,860$; $\$6,860 \div 100 = \$68.60$
Add: $\$980 + \$68.60 = \$1,048.60$

8. **16%** Subtract: $\$5,000 - \$4,200 = \$800$
$\frac{\text{amount of change}}{\text{original amount}} = \frac{\text{rate}}{100}$ $\frac{\$800}{\$5,000} = \frac{x}{100}$
$\$800 \times 100 = \$80,000$; $\$80,000 \div \$5,000 = 16$;
$\frac{16}{100} = 16\%$

GED Math Practice • Using Proportions to Solve Problems (page 115)

1. **(5)** $\frac{500 \times 7}{3}$ Set up the proportion: $\frac{3}{500} = \frac{7}{x}$
Cross multiply and divide by the remaining term:
$500 \times 7 \div 3$. Rewrite to match an answer choice:
$500 \times 7 \div 3 = \frac{500 \times 7}{3}$

2. **(4) \$1,800** Set up the proportion: $\frac{\$300}{5} = \frac{x}{30}$
Cross multiply and divide by the remaining term:
$\$300 \times 30 \div 5 = \$1,800$

3. **(3)** $\frac{\$0.28 \times \$92,000}{\$1,000}$
Set up the proportion: $\frac{\$0.28}{\$1,000} = \frac{x}{\$92,000}$
Cross multiply and divide by the remaining term:
$\$0.28 \times \$92,000 \div \$1,000$.
Rewrite to match an answer choice: $\frac{\$0.28 \times \$92,000}{\$1,000}$

4. **(2)** $\frac{1.5 \times 2}{3}$ Set up the proportion: $\frac{1.5}{3} = \frac{x}{2}$
Cross multiply and divide by the remaining term:
$1.5 \times 2 \div 3$. Rewrite to match an answer choice:
$1.5 \times 2 \div 3 = \frac{1.5 \times 2}{3}$

5. **(1)** $\frac{65 \times 3}{1}$ Set up the proportion: $\frac{65}{1} = \frac{x}{3}$
Cross multiply and divide by the remaining term:
$65 \times 3 \div 1$. Rewrite to match an answer choice:
$65 \times 3 \div 1 = \frac{65 \times 3}{1}$

6. **(3) \$15.00** Set up the proportion: $\frac{3}{\$7.50} = \frac{6}{x}$
Cross multiply and divide by the remaining term:
$\$7.50 \times 6 \div 3 = \$15.00$

GED Math Practice • Percents on the Calculator and Grid (page 117)

1. **35%**

2. **6.25%**

3. **13,500**

4. **140**

5. **2,000**

6. **27.5**

7. **(1) 6%** Set up the proportion: $\frac{\$16.74}{\$279} = \frac{x}{100}$
Cross multiply and divide by the remaining term:
$\$16.74 \times 100 \div \$279 = 6$, and $\frac{6}{100} = 6\%$

8. **.49** First add the prices:
$\$3.49 + \$1.99 + \$1.99 = \$7.47$.
Use proportion to find the amount
of sales tax: $\frac{x}{\$7.47} = \frac{6.5}{100}$
Cross multiply and divide by
the remaining term: $\$7.47 \times 6.5 \div$
$100 = \$0.4855$, which rounds to
$\$0.49$ or $\$.49$

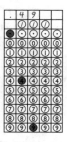

9. **5.87** "15% off" means the sale price is 85% of the original price. Use proportion to find the original price: $\frac{\$4.99}{x} = \frac{85}{100}$ Cross multiply and divide by the remaining term: $\$4.99 \times 100 \div 85 = \$5.87$

GED Math Review
(pages 118–119)
Part I

1. **(1) $11,250** Set up a proportion: $\frac{x}{45,000} = \frac{25}{100}$ Cross multiply and divide by the remaining term: $\$45,000 \times 25 \div 100 = \$11,250.$

2. **(1) 10%** First find the amount of increase in rent: $\$14,850 - \$13,500 = \$1,350$. Then set up a proportion: $\frac{\$1,350}{\$13,500} = \frac{x}{100}$ Cross multiply and divide by the remaining term: $\$1,350 \times 100 \div \$13,500 = 10$, and $\frac{10}{100} = 10\%$

3. **(5) $490** Set up a proportion: $\frac{\$350}{\$5,000} = \frac{x}{\$7,000}$ Each fraction represents the ratio of the amount spent on awards to the total budget. Cross multiply and divide by the remaining term: $\$350 \times \$7,000 \div \$5,000 = \$490$

4. **(4) $1,428** Set up a proportion: $\frac{8}{\$68} = \frac{168}{x}$ Cross multiply and divide by the remaining term: $\$68 \times 168 \div 8 = \$1,428$

5. **24.5 or $\frac{49}{2}$** Write a proportion: $\frac{5}{7} = \frac{17.5}{x}$ Each fraction represents the ratio of the height to the width of the picture. Cross multiply and divide by the remaining term: $7 \times 17.5 \div 5 = 24.5$ or $\frac{49}{2}$

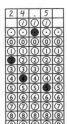

Part II

6. **(1) 20%** Since the problem asks for the percent of the total pay, first find the sum of the deductions and her take-home pay. $\$120 + \$480 = \$600$. Then set up a proportion: $\frac{\$120}{\$600} = \frac{x}{100}$ Cross multiply and divide by the remaining term: $\$120 \times 100 \div \$600 = 20$, and $\frac{20}{100} = 20\%$

7. **(4) $800** Find the amount of interest for one year: $\frac{x}{\$5,000} = \frac{8}{100}$ Cross multiply and divide by the remaining term: $\$5,000 \times 8 \div 100 = \$400$. Multiply the interest for one year by two, since he will pay it back in 2 years: $\$400 \times 2 = \$800.$

8. **(4) 12.5%** First find the discount: $\$80 - \$70 = \$10$. To find the percent decrease, set up a proportion comparing the discount to the original price: $\frac{\$10}{\$80} = \frac{x}{100}$ Cross multiply and divide by the remaining term: $\$10 \times 100 \div \$80 = 12.5$, and $\frac{12.5}{100} = 12.5\%$

9. **(2) $\frac{1 \times 16}{4}$** Write a proportion: $\frac{1}{4} = \frac{x}{16}$ Each fraction represents the ratio of the amount of sugar to the amount of water. Cross multiply and divide by the remaining term: $1 \times 16 \div 4 = \frac{1 \times 16}{4}$

---

10. **25** Set up a proportion: $\frac{2}{20} = \frac{2.5}{x}$ Cross multiply and divide by the remaining term: $20 \times 2.5 \div 2 = 25$. The towns are 25 miles apart.

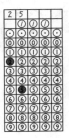

PROGRAM 25: MEASUREMENT

Practice 1 (page 123)
A.
1. 15 weeks
2. 24 cups
3. 45 inches
4. 4 cups
5. 108 inches
6. 10,560 feet
7. 2.5 or $2\frac{1}{2}$ years
8. 72 ounces
9. 1 pt. = 2 c. = **16 fluid ounces**

B.
10. (2) ounces
11. (2) cups
12. (2) gallons
13. (1) inches

Practice 2 (page 125)
A.
1. 1.5 kilograms
2. 2,400 meters
3. 3,700 millimeters
4. 3.25 kilometers
5. 400 centimeters
6. 4 kiloliters
7. 180 millimeters
8. 12,250 grams

B.
9. (2) grams
10. (3) meters

Practice 3 (page 127)
A.
1. 11 ft. 4 in.
2. 50 min.
3. 4.4 L
4. 5.46 m
5. 15 lb. 13 oz.
6. 1 yd. 1 ft. 8 in.
7. **7.15 kg** or **7,150 g** To find the sum in kilograms, convert grams to kilograms: 1,550 g = 1.55 kg. Then add: 5.6 kg + 1.55 kg = 7.15 kg. To find the sum in grams, convert kilograms to grams: 5.6 kg = 5,600 g. Add to find the total: 5,600 g + 1,550 g = 7,150 g.
8. **52 cm** 1 m = 100 cm 100 cm − 48 cm = 52 cm

B.
9. **4 hr. 20 min.** Add the hours: 1 + 1 = 2 hr. Add the minutes: 45 + 20 + 30 + 45 = 140 min. Change 140 minutes to hours and minutes: 140 min. = 2 hr. 20 min. Find the total: 2 hr. + 2 hr. 20 min. = 4 hr. 20 min.
10. **5 pt.** Use the facts 1 qt. = 2 pt. and 1 gal. = 4 qt. From this, you can determine that 1 gal. = 4 qt. = 8 pt. Subtract to find the amount remaining: 8 pt. − 3 pt. = 5 pt.
11. **27 ft. 9 in.** Find the total he used on the screens: 4 ft. 6 in. + 3 ft. 9 in. = 7 ft. 15 in. = 8 ft. 3 in. You know that 1 yd. = 3 ft., so 12 yd. = 36 ft. Subtract the total used from 36 feet: 36 ft. − 8 ft. 3 in. = 27 ft. 9 in.
12. **700 g** 2.8 kg = 2,800 g Subtract: 3,500 g − 2,800 g = 700 g
13. **9.415 kg** Add: 3.64 + 5.4 + 0.375 = 9.415 kg

14. **1.45 m** 35 cm = 0.35 m
Subtract: 1.8 m − 0.35 m = 1.45 m

15. **8 lb. 7 oz.** Subtract: 20 lb. − 11 lb. 9 oz. =
19 lb. 16 oz. − 11 lb. 9 oz. = 8 lb. 7 oz.

Practice 4 (page 129)

A. **1.** 5 lb.   **5.** 1 gal. 3 qt.
2. 6 in.   **6.** 6 lb.
3. 45 kg   **7.** 0.6 L
4. 2.52 m   **8.** 2.4 m

B. **9.** **5 lb.** 10 oz.× 8 = 80 oz. Use the fact
1 lb. = 16 oz. 80 oz. ÷ 16 = 5 lb.

10. **10 hr. 40 min.** 1 hr. 20 min. × 8 = 8 hr. 160
min. Change 160 min. to hours using the fact 1
hr. = 60 min. 160 min. = 2 hr. 40 min. His total
overtime is 8 hr. + 2 hr. 40 min. = 10 hr. 40 min.

11. **1.5 L** 7.5 L ÷ 5 = 1.5 L

12. **7.8 kg** 1.3 kg × 6 = 7.8 kg

13. **15 ft.** For each length of lacing, find the
amount needed. Multiply:
2 ft. 3 in. × 4 = 8 ft. 12 in. = 9 ft.
Multiply: 9 in.× 8 = 72 in. = 6 ft.
Add: 9 + 6 = 15 ft.

14. **16 paper cups** Find out how many ounces
are in 1 gallon. 1 gal. = 4 qt. = 8 pt. = 16 c. =
128 oz. Divide by 8: 128 ÷ 8 = 16 paper cups.

Practice 5 (page 131)

A. **1.** perimeter: 2.5 × 4 = **10 centimeters**
area: 2.5 × 2.5 = **6.25 square centimeters**

2. perimeter: (5 × 2) + (3 × 2) = **16 feet**
area: 5 × 3 = **15 square feet**

3. perimeter: (7.8 × 2) + (2 × 2) = **19.6 centimeters**
area: 7.8 × 2 = **15.6 square centimeters**

B. **4.** 11 × 3 × 7 = **231 cubic centimeters**

5. 2 × 2.5 × 8 = **40 cubic feet**

6. 5 × 5 × 5 = **125 cubic inches**

C. **7.** **50 ft.** Find the perimeter:
(13 × 2) + (12 × 2) = 26 + 24 = 50 ft.

8. **140 square feet**
Find the volume: 7 × 4 × 5 = 140 sq. ft.

Practice 6 (page 133)

1. **222 sq. ft.** Divide the kitchen into two rectangles.
One is 15 by 7 ft. The other is 13 by 9 ft. Find the
areas: 15 × 7 = 105 sq. ft.; 13 × 9 = 117 sq. ft.
Add: 105 + 117 = 222 square feet.

2. **11 cubic feet** Break the slab into two rectangular
sections. One is 5 by 2 ft.; the other is 6 by 2 ft.
Find the volume of each section. The thickness of
the slab is $\frac{1}{2}$ ft. 5 × 2 × $\frac{1}{2}$ = 5 cubic feet, and
6 × 2 × $\frac{1}{2}$ = 6 cubic feet. Add: 5 + 6 = 11 cubic feet.

3. **20 ft.** The measures of the unmarked sides are
2 ft. and 1 ft. Add all the sides: 5 + 2 + 1 + 2 + 6 + 4
= 20 ft.

4. **185 square meters** Find the area of the entire
room: 18 × 12.5 = 225 square meters. Find the
area of the pool: 8 × 5 = 40 square meters.
Subtract the area of the pool from the area of the
entire room: 225 − 40 = 185 square meters.

5. **1,120 cubic inches** The box can be divided into
two rectangular containers. One measures 8 by 8
by 5 inches. The other measures 20 by 8 by 5
inches. Find the volumes: 8 × 8 × 5 = 320 cubic
inches, and 20 × 8 × 5 = 800 cubic inches.
Add: 320 + 800 = 1,120 cubic inches.

GED Math Practice • Draw a Picture (page 135)

1. **50 feet** Draw and label a
picture of the deck. Add to
find the distance around the
3 sides of the deck: 15 feet +
15 feet + 20 feet = 50 feet
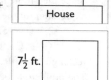

2. **240 square feet** Draw and
label one of the walls, since
all four walls are the same.
Multiply the length by the
height: 8 feet × 7.5 feet = 60
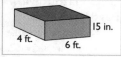
square feet. Then multiply by the number of walls:
60 sq. ft.× 4= 240 square feet.

3. **30 cubic feet** Draw
and label a picture of
the pool. Change 15
inches to feet: $\frac{15}{12}$ =

1.25. Multiply the length, width, and height:
6 × 4 × 1.25 = 30 cubic feet

4. **2 × (3 + 4 + 5)** Draw and label
a picture of one of the windows.
Write an expression adding the
lengths of the sides: 3 + 4 + 5.
Since there are two windows,

multiply the expression by 2: 2 × (3 + 4 + 5)

5. **112 feet** Draw and label a
picture of the park. Find the
perimeter of the picnic area:
4 × 30 feet = 120 feet.
Subtract the widths of the two
openings: 120 feet − 2 × 4
feet = 112 feet.

6. **40 blocks** Draw and
label a picture of the
quilt. Find the
measurements of the
area inside the border.
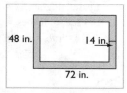
Subtract twice the width
of the border from each dimension: 48 − 8 = 40
inches and 72 − 8 = 64 inches. Divide each
dimension by the width or length of the block: $\frac{40}{8}$ = 5
blocks for the width, and $\frac{64}{8}$ = 8 blocks for the
length. Multiply: 5 × 8 = 40 blocks.

GED Math Practice • Formulas Page (page 137)

1. **$1.27** Use the cost formula, c = nr. Substitute:
$7.59 = 6 × r. Solve for r: r = $\frac{\$7.59}{6}$ = $1.27.

2. **5.25 square feet** Use the area formula for a
triangle, A = $\frac{1}{2}$bh. Substitute and solve:
A = $\frac{1}{2}$ × 3 feet × 3.5 feet = 5.25 square feet.

3. **18 inches** Use the formula for volume of a rectangular container, $V = lwh$. Change the inches to feet: 24 inches = 2 feet, and 15 inches = 1.25 feet. Substitute: 3.75 cubic feet = 2 feet × 1.25 feet × h. Solve for h: $h = \frac{3.75}{2 \times 1.25} = \frac{3.75}{2.5} = 1.5$ feet. Change 1.5 feet to inches: $1.5 \times 12 = 18$ inches.

4. **15.7 inches** Use the circumference formula for a circle, $C = \pi d$. Substitute using $\pi = 3.14$ and solve: $C = 3.14 \times 5$ inches = 15.7 inches.

5. **(3) \$2,500 + (\$2,500 × .07 × 2)** Use the interest formula, $i = prt$. Substitute: $i = \$2,500 \times .07 \times 2$. Since the problem asks for the total, add the expression for the interest to the amount of the loan: \$2,500 + (\$2,500 × .07 × 2).

6. **(1) $\frac{8.5 \times 11}{144}$** Use the formula for area of a rectangle, $A = lw$. Substitute: $A = 8.5 \times 11$. The problem asks for the answer in square feet. There are 144 square inches in a square foot: 12 inches × 12 inches = 144 square inches. Divide the expression for the area by 144: $\frac{8.5 \times 11}{144}$.

GED Math Review (pages 138–139)
Part I

1. **(4) 4.25%** Use the interest formula, $i = prt$. Substitute: $\$85 = \$2,000 \times r \times 1$. Solve for r: $r = \frac{\$85}{\$2,000} = 0.0425$. Change to a percent: 0.0425 = 4.25%.

2. **(2) 2** Use the formula for volume of a rectangular container, $V = lwh$. Substitute: 6 cubic feet = 2 feet × 1.5 feet × h. Solve for h: $h = \frac{6}{2 \times 1.5} = 2$ feet.

3. **(1) 314,000** Use the formula for volume of a cylinder, $V = \pi r^2 h$. Divide the diameter by 2 since the formula uses the radius: $\frac{100}{2} = 50$. Substitute and solve: $V = 3.14 \times 50$ cm × 50 cm × 40 cm = 314,000 cubic centimeters.

4. **(3) 2,123** Use the formula for the area of a circle, $A = \pi r^2$. Divide the diameter by 2 since the formula uses the radius: $\frac{52}{2} = 26$. Substitute using $\pi = 3.14$ and solve: $A = 3.14 \times 26 \times 26 = 2,123$ feet rounded to the nearest square foot.

5. **12** Use the formula for the area of a parallelogram, $A = bh$. Substitute: 144 square inches = 12 inches × h. Solve for h: $h = \frac{144}{12} = 12$ inches.

Part II

6. **(4) $\frac{200}{3}$** Use the distance formula, $d = rt$. Substitute: $200 = r \times 3$. Solve for r: $r = \frac{200}{3}$.

7. **(2) 26.4** Use the distance formula, $d = rt$. Change miles to feet: $\frac{1}{4}$ mile = $\frac{5,280}{4}$ = 1,320 feet. Substitute: $1,320 = r \times 50$. Solve for r: $r = \frac{1,320}{50}$ = 26.4 feet per second.

8. **(2) 157.0** Use the formula for the circumference of a circle, $C = \pi d$. Since the formula uses the diameter, multiply the radius by 2: 2 × 25 cm = 50 cm. Substitute and solve: $C = 3.14 \times 50$ cm = 157 cm.

9. **(5) 18** Draw and label a picture of the box. Draw and label a picture of the shelf. Notice that the shelf and the boxes are both 12 inches deep so you only need to work with the length and the height. Each box is 1 foot (or 12 inches) wide, so six boxes will fit across the shelf. Three boxes will fit on top of each other. Multiply: 3 × 6 = 18 boxes.

10. **(5) $\frac{2.99}{12}$** Use the total cost formula, $c = nr$. Substitute: $\$2.99 = 12 \times r$. Solve for r: $r = \frac{2.99}{12}$.

11. **26** Add the lengths of the sides of the figure: 3 + 7 + 5 + 4 + 7 = 26 inches.

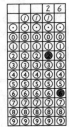

PROGRAM 26:
DATA ANALYSIS

Practice 1 (page 143)

1. XON
2. \$8.53
3. \$2,817.60
4. \$5.78
5. \$2.50
6. **72 workers** Add the symbols in the rows labeled "Fewer than 5" and "5.1 to 10." 3 + 6 = 9. Then multiply by the number of workers per symbol. 9 × 8 = 72 workers
7. **16 workers** There are 2 more symbols in the third row than the fourth. 2 × 8 = 16 workers
8. 27 feet
9. The Rocket

Practice 2 (page 145)

Your answers will vary slightly if you used different values for the bars.

1. **about 350 customers**
2. **about 600 customers** 350 + 250 = 600
3. **about 25 customers** 250 − 225 = 25
4. about 70,000 claims
5. March and May
6. 40,000 claims
7. 60% increase
8. FunCube
9. **80,000** 25 + 15 + 40 = 80 thousand
10. **GameBox** 50 to 24 is close to 2 to 1.
11. ocean theme
12. **1:4** 20:80 = 1:4
13. **75%** First find the total sold: 10 + 20 + 90 = 120 dinosaur shirts. Then find the percent: 90 is 75% of 120.
14. **170 small/medium shirts** 30 + 50 + 90 = 170
15. **jungle theme** You don't need to add the numbers. Imagine combining the lengths of the three bars for each theme. The total length of the jungle bars would be less than the length of the longest dinosaur bar and less than the length of any two of the ocean bars.

Practice 3 (page 147)

1. 2001
2. 2002
3. 1998
4. 5 hours
5. week 3
6. week 4

7. **2 hours** Except for a small increase one week, Aaron's practice time has stayed the same or decreased each week. Although it is possible that his practice time could make a sudden increase, the best prediction is that it will continue to decrease.

8. August

9. No. The sales of stuffed-crust pizza have been steadily increasing, and nothing in the data suggests the sales are slowing. If the trend continues, sales of stuffed-crust pizza will catch up to and pass the sales of Chicago-style pizza in the coming months.

10. May
11. April
12. April
13. June

Practice 4 (page 149)

1. **Hayes** $\frac{3}{10} = \frac{30}{100} = 30\%$. The amount closest to 30% is Hayes' 29%.

2. **95%**

3. **75%** Add the percents for Perez and Hayes: $46\% + 29\% = 75\%$.

4. **$1,900**

5. **Housing and Transportation** $50\% = \frac{1}{2}$. 50% or $\frac{1}{2}$ of $2,400 is $1,200. Look for two items with a total budget of $1,200.

6. $\frac{1}{6}$ Write a fraction and reduce: $\frac{\$400}{\$2,400} = \frac{1}{6}$.

7. **$3,200** For Housing to be $\frac{1}{4}$ of the total budget, the Pace family would have to earn four times the Housing budget of $800. $800 \times 4 = \$3,200$

8. **food and vet services** One-third of a dollar is about 33 cents. Food and Vet Services total 32 cents.

9. **$116** Solve the proportion: $\frac{\$0.58}{\$1} = \frac{x}{\$200}$.

10. **4%** 4 cents out of 100 cents is $\frac{4}{100} = 4\%$.

11. **$25** Find 5% of $500. $\frac{x}{\$500} = \frac{5}{100}$

12. **public and private bonds** $\frac{3}{4} = 75\%$. Find the two items that total 75%.

13. **$800** Find 20% of $4,000. $\frac{x}{\$4,000} = \frac{20}{100}$

14. **70%** Subtract 30%, the amount invested in private bonds, from 100%.

Practice 5 (page 151)

1. 465 miles
2. a. 83
   b. 78
   c. The median best represents his normal score because four of the scores were in the 70s. The high score of 109 is having too great an effect on the mean.
3. 3T
4. **69°** The two middle temperatures are 68° and 70°. Find the mean: $68° + 70° = 138°$. Divide by 2: $138° \div 2 = 69°$.
5. $104.38

6. 4.4 inches
7. February
8. The **mean** would change more because April's measurement is already the lowest in the list. Making it lower will not change the median at all. However, the lower number will be farther from the center, so it will have a great effect on the mean.

Practice 6 (page 153)

1. $\frac{3}{7}$

2. $\frac{4}{80} = \frac{1}{20}$

3. $\frac{6}{8} = \frac{3}{4}$

4. $\frac{5}{20} = 25\%$

5. a. $\frac{4}{20} = 20\%$
   b. 20% of 50 = **10 cards**

6. $\frac{15}{25} = 60\%$ If 10 out of 25 are doctors, then 15 out of 25 are not doctors.

GED Math Practice • The Problem-Solving Method (page 155)

1. Question asks: for the difference in sales for January and June
   Facts needed: $3,000 and $1,000
   Operation: subtract
   Set up and solve: $3,000 - \$1,000 = \$2,000$
   Check: $2,000 + \$1,000 = \$3,000$

2. Question asks: for the average monthly sales for May and June
   Facts needed: $1,750 and $1,000
   Operations: add and divide the sum by the number of months
   Set up and solve: $\frac{\$1,750 + \$1,000}{2} = \$1,375$
   Check: $\frac{\$1,750 + \$1,000}{2} = \$1,375$

3. Question asks: for the number of employees with fewer than 11 years of education
   Facts needed: 50 employees, 8%
   Operation: multiply
   Set up and solve: $8\% = .08$, and $50 \times .08 = 4$
   Check: $50 \times .08 = 4$

4. Question asks: what percent of employees have 12 or more years of education
   Facts needed: 52% and 16%
   Operation: add
   Set up and solve: $52\% + 16\% = 68\%$
   Check: $68\% - 16\% = 52\%$

GED Math Practice • Data Analysis on the Calculator and Grid (page 157)

1. 9.2
2. 20
3. 4.85
4. $265
5. **(3) $82,500** To find the mean, add the dollar amounts and divide by 4:
   $\frac{\$75,000 + \$82,500 + \$77,500 + \$95,000}{4} =$
   $\frac{\$330,000}{4} = \$82,500$.

6. 266 Multiply the mean by 4 to find the total amount of money Tamra could earn in 4 weeks: $250 × 4 = $1,000. Subtract the sum of the amounts given for the three weeks from the total: $1,000 − ($237 + $245 + $252) = $1,000 − $734 = $266.

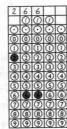

GED Math Review (pages 158–159)
Part I

1. **(1) $250** Find the difference in the taxes for 1997 and 2001. Read the amount of taxes for each year from the graph: 1997 ($1,250) and 2001 ($1,500). Subtract: $1,500 − $1,250 = $250.

2. **(3) $1,350** Add the amounts and divide by the number of years:
$$\frac{\$1,250 + \$1,300 + \$1,300 + \$1,400 + \$1,500}{5} =$$
$$\frac{\$6,750}{5} = \$1,350.$$

3. **(4) $\frac{2}{5}$** Four of the ten socks are red, so the probability of selecting a red sock is $\frac{4}{10}$, or $\frac{2}{5}$.

4. **(5) 55** The question asks for the total enrollment for the first year. The bars for the first year show 25 girls and 30 boys. Add: 25 + 30 = 55.

5. **10** The question asks for the difference in the enrollment for boys and girls in the fourth year. The bars for the fourth year show 75 girls and 65 boys. Subtract: 75 − 65 = 10.

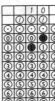

Part II

6. **(2) $1,250** The question asks for the total amount Lea has in her budget for the month. Add the amounts for each section of the graph: $500 + $300 + $250 + $100 + $100 = $1,250.

7. **(4) 40%** To find the percent, divide the dollars paid for rent by the total dollars:
$\frac{\$500}{\$1,250} = 0.4 = \frac{40}{100}$, or 40%.

8. **(3) 52.4** Add the numbers and divide by 5:
$\frac{50 + 40 + 65 + 54 + 53}{5} = \frac{262}{5} = 52.4.$

9. **(3) $\frac{1}{2}$** Three of the six sections on the spinner have even numbers. The probability is $\frac{3}{6}$, or $\frac{1}{2}$.

10. **2/3** Four of the six sections have numbers greater than two. The probability is 4/6, or 2/3.

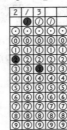

EXTRA PRACTICE PAGES
Place Value (page 160)

A.
1. 4 × 10 = **40**
2. 1 × 1 = **1**
3. 6 × 1,000 = **6,000**
4. 0 × 1,000 = **0**
5. 7 × 10 = **70**
6. 6 × 100 = **600**
7. 9 × 10,000 = **90,000**
8. 9 × 100,000 = **900,000**
9. 0 × 10,000 = **0**
10. 1 × 1,000,000 = **1,000,000**

B.
11. one hundred ninety-two
12. three hundred forty-seven
13. four thousand, six
14. seven thousand, two hundred thirty
15. thirty-four thousand, eight hundred fifty-one
16. seventy thousand, five hundred
17. three hundred twenty-three thousand
18. five hundred sixty thousand, seven hundred
19. four million, two hundred thousand
20. seven million, four hundred one

C.
21. 376
22. 208
23. 930
24. 8,024
25. 1,005
26. 60,900
27. 95,024
28. 300,050
29. 5,200,000
30. 1,000,008

Comparing and Ordering (page 161)

A.
1. <
2. >
3. =
4. >
5. <
6. <
7. =
8. >
9. >
10. >
11. <
12. >

B.
13. 140, 145, 149
14. 3,560, 3,980, 4,540
15. 34,954, 34,945, 34,495
16. 5,871, 5,081, 581

C.
17. $2,450
18. House B
19. House D
20. $99,200, $98,900, $95,800, $95,700
21. Ontario, Erie, Michigan, Huron, Superior
22. 25,807, 25,061, 24,942
23. the second quarter

Number Patterns (page 162)

A.
1. **22** (add 2)
2. **17** (subtract 4)
3. **55** (subtract 10)
4. **27, 36** (add 9)
5. **40, 55** (add5)
6. **50** (subtract 20)
7. **64, 80** (add 8)
8. **425** (subtract 100)

B.
9.
10.

C. **11.** 30 pounds

12. **1,330** Subtract each previous year's population from the following year's population to see the pattern. $1,240 - 1,230 = 10$
$$1,260 - 1,240 = 20$$
$$1,290 - 1,260 = 30$$
If the pattern continues, next year's population will be 40 more than this year.
$$1,290 + 40 = 1,330$$

13. 6, 12, 18, 24

14. 64

15. May 31

16. 14, 28, 42

Calculator Basics (page 163)

A. **1.** c   **5.** b
2. f   **6.** g
3. d   **7.** e
4. a

B. **8.** 503.   **12.** 324589.
9. 9437.   **13.** 6598200.
10. 20461.   **14.** 98003736.
11. 610.   **15.** 800700.

C. **16.** 245.   **19.** 849877.
17. 5009.   **20.** 30560.
18. 6372.

Adding and Subtracting Whole Numbers (page 164)

A. **1.** 69   **5.** 8,904
2. 52   **6.** 3,320
3. 378   **7.** 704
4. 1,879   **8.** 7,067

B. **9.** 83   **11.** 3,875
10. 1,066   **12.** 2,000

C. **13.** 94   **16.** 16
14. 29   **17.** 76
15. 122   **18.** 120

D. **19.** **$1,197** $180 + $352 + $256 + $409 = $1,197
20. **$11,300** $27,100 - $25,800 = $1,300
21. **2,317** $5,678 - 3,361 = 2,317$
22. **6,681** $4,249 + 2,432 = 6,681$

Multiplying and Dividing Whole Numbers (page 165)

A. **1.** 207   **6.** 76
2. 22   **7.** 8,032
3. 1,295   **8.** 407 r 1
4. 402   **9.** 15,000
5. 1,230

B. **10.** 25   **13.** 46,200
11. 1,312   **14.** 58 r 29
12. 25   **15.** 211,800

C. **16.** **135 miles** $27 \times 5 = 135$
17. **$21,360** $1,780 \times 12 = $21,360
18. **$375** $4,500 \div 12 = $375
19. **67 gallons** $2,345 \div 35 = 67$

20. **16 children** $128 \div 8 = 16$
21. **6 inches per month** $72 \div 12 = 6$

Estimating (page 166)

A. All reasonable estimates are acceptable.

	Estimate	Exact Answer
1.	$2,300 + 13,000 = \mathbf{15,300}$	**15,237**
2.	$4,600 - 3,000 = \mathbf{1,600}$	**1,523**
3.	$400 \times 30 = \mathbf{12,000}$	**12,059**
4.	$5,300 \div 10 = \mathbf{530}$	**445 r 1**
B. **5.**	$\$4,000 + \$7,000 = \mathbf{\$11,000}$	**$11,999**
6.	$500,000 - 200,000 = \mathbf{300,000}$	**230,493**
7.	$1,000 \times 30 = \mathbf{30,000}$	**73,242**
8.	$20,000 \div 20 = \mathbf{1,000}$	**951 r 22**
C. **9.**	$400 \div 20 = \mathbf{20}$	**18 r 9**
10.	$8,000 \div 40 = \mathbf{200}$	**202 r 14**
11.	$10,000 \div 50 = \mathbf{200}$	**193 r 13**

D. All reasonable estimates are acceptable.

12. $\$70 + \$50 + \$140 = \mathbf{\$260}$
$\$72 + \$45 + \$137 = \mathbf{\$254}$

13. $\$350 - \$250 = \mathbf{\$100}$
$\$350 - \$259 = \mathbf{\$91}$

14. $20 \times 10 = \mathbf{200}$
$11 \times 18 = \mathbf{198}$

15. $\$12,000 \div 4 = \mathbf{\$3,000}$
$\$12,416 \div 4 = \mathbf{\$3,104}$

Calculator Operations and Grid Basics (page 167)

A. **1.** 974   **6.** 157
2. 24   **7.** 2,126
3. 18,270   **8.** 422
4. 19   **9.** 7,680
5. 372   **10.** 3,608

B. **11.** **88** Find the total amount to be paid: $480 + $48 = $528. Divide to find the amount of each payment:
$528 \div 6 = $88.

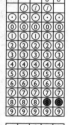

12. **875** Multiply to find the total number of miles:
$175 \times 5 = 875$ miles.

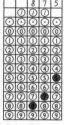

13. **13440** First find the take-home pay for each month:
$\$1,400 - \$280 = \$1,120$. Then multiply by 12 to find the yearly take-home pay:
$\$1,120 \times 12 = \$13,440$.

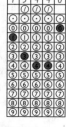

14. **10700** Add to find the total: $3,500 + $2,300 + $4,900 = $10,700.

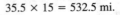

Rounding and Comparing Decimals (page 168)

A.
1. $1.50
2. 36.6
3. 9.37
4. 2
5. 27.62
6. $14
7. $0.56
8. 0.1

B.
9. >
10. >
11. =
12. >
13. <
14. >
15. >
16. <
17. =
18. <

C.
19. $32.50, $31.55, $31.25
20. 900 milliliters
21. almonds, pecans, peanuts, cashews
22. $1.41
23. 62.4 lb.
24. **2.79 in.** 2.7844 rounds to 2.78 inches, and 2.7861 rounds to 2.79 inches

Adding and Subtracting Decimals (page 169)

A.
1. 8.1
2. 1.2
3. 2.70
4. $50.00
5. 3.716
6. 6.117

B.
7. 10.152
8. $0.04
9. 9.282
10. 18.748
11. 10.627
12. 28.51

C.
13. $86.20
14. 25.6 feet
15. $7.06
16. $15.81
17. 31.4 degrees
18. 14.6 m

Multiplying and Dividing Decimals (page 170)

A.
1. 34.2
2. 4.3
3. 0.036
4. $8.25
5. $25.20
6. 37.0
7. 51.0
8. 20.2
9. **$3.20** $3.1960 rounds to $3.20

B.
10. 9.9
11. 0.0084
12. $0.60
13. **$14.66** $14.6625 rounds to $14.66
14. **$0.42** $0.416 rounds to $0.42
15. 1.044

C.
16. $1,228.50
17. **$11.24** $11.235 rounds to $11.24
18. $177.50
19. $682.50
20. $1.50
21. 25 miles per gallon

Decimal Calculator Operations and Grids (page 171)

A.
1. 17.47
2. $83.75
3. 34.65
4. 16.98
5. 15.51
6. 7.27
7. 3.01
8. $0.7616
9. $7.11
10. 21.85

B. 11. **532.5** Multiply the number of miles per gallon by the number of gallons: $35.5 \times 15 = 532.5$ mi.

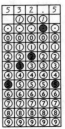

12. **1.25** Divide the total pounds by the number of parts: 15 lb. ÷ 12 = 1.25 lb.

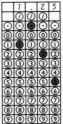

13. **56.1** Add the temperatures and divide the sum by 5: 57.2 + 59.0 + 54.5 + 60.1 + 49.7 = 280.5, and 280.5 ÷ 5 = 56.1 degrees.

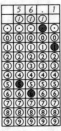

14. **2.4** Subtract to find the difference: 148.2 − 145.8 = 2.4 lb.

Fraction Basics (page 172)

A.
1. $\frac{1}{2}$
2. $\frac{8}{9}$
3. $\frac{1}{4}$
4. $\frac{2}{5}$
5. $\frac{1}{6}$
6. $\frac{1}{4}$

B.
7. $\frac{6}{12}$
8. $\frac{8}{20}$
9. $\frac{9}{24}$
10. $\frac{16}{36}$
11. $\frac{12}{21}$
12. $\frac{30}{50}$

C.
13. >
14. >
15. =
16. =
17. <
18. <

D.
19. $\frac{7}{10}$ $\frac{35}{50} = \frac{7}{10}$
20. $\frac{1}{3}$ $\frac{30}{90} = \frac{1}{3}$
21. $\frac{2}{9}$ $\frac{$100}{$450} = \frac{10}{45} = \frac{2}{9}$
22. $\frac{5}{8}$ $\frac{25,000}{40,000} = \frac{25}{40} = \frac{5}{8}$

Adding and Subtracting Fractions (page 173)

A.
1. $\frac{4}{5}$
2. $\frac{4}{8} = \frac{1}{2}$
3. $\frac{7}{7} = 1$
4. $\frac{4}{10} = \frac{2}{5}$
5. $\frac{7}{12}$
6. $\frac{7}{24}$
7. $14\frac{4}{5}$
8. $7\frac{4}{6} = 7\frac{2}{3}$
9. $13\frac{4}{8} = 13\frac{1}{2}$
10. $9\frac{2}{7}$
11. $41\frac{8}{10} = 41\frac{4}{5}$
12. $1\frac{7}{24}$
13. $7\frac{13}{8} = 8\frac{5}{8}$
14. $9\frac{9}{20}$
15. $8\frac{22}{15} = 9\frac{7}{15}$
16. $10\frac{1}{12}$
17. $15\frac{2}{9}$
18. $4\frac{7}{12}$
19. $10\frac{41}{30} = 11\frac{11}{30}$
20. $\frac{14}{15}$
21. $21\frac{9}{8} = 22\frac{1}{8}$
22. $6\frac{17}{30}$
23. $40\frac{19}{16} = 41\frac{3}{16}$
24. $9\frac{5}{12}$

B.
25. $1\frac{5}{8}$ yd.
26. $4\frac{1}{2}$ yd.
27. $9\frac{1}{2}$ hr.
28. $2\frac{1}{4}$ acres

Multiplying and Dividing Fractions (page 174)

A.
1. $\frac{1}{6}$
2. $\frac{3}{4}$
3. $\frac{1}{8}$
4. $\frac{9}{4} = 2\frac{1}{4}$
5. $\frac{4}{15}$
6. $\frac{1}{12}$
7. 36
8. $\frac{20}{9} = 2\frac{2}{9}$
9. $\frac{49}{8} = 6\frac{1}{8}$
10. $\frac{70}{11} = 6\frac{4}{11}$
11. $\frac{35}{6} = 5\frac{5}{6}$
12. $\frac{11}{3} = 3\frac{2}{3}$
13. $\frac{19}{4} = 4\frac{3}{4}$
14. $\frac{13}{4} = 3\frac{1}{4}$
15. $\frac{51}{4} = 12\frac{3}{4}$
16. $\frac{29}{2} = 14\frac{1}{2}$
17. $\frac{88}{9} = 9\frac{7}{9}$
18. 35

B.
19. **$26** $4\frac{1}{3} \times \$6 = \frac{13}{3} \times \frac{\$6}{1} = \$26$
20. **18 miles** $117 \div 6\frac{1}{2} = \frac{117}{1} \times \frac{2}{13} = 18$ mi.
21. **$11\frac{1}{4}$ ounces** $3\frac{3}{4} \times 3 = \frac{15}{4} \times 3 = \frac{45}{4} = 11\frac{1}{4}$ oz.
22. **78 people** $26 \div \frac{1}{3} = 26 \times 3 = 78$ people
23. **20 lengths** $315 \div 15\frac{3}{4} = \frac{315}{1} \div \frac{63}{4} = \frac{315}{1} \times \frac{4}{63} = 20$ lengths
24. **6 boards** $33 \div 5\frac{1}{2} = \frac{33}{1} \div \frac{11}{2} = \frac{33}{1} \times \frac{2}{11} = 6$ boards
25. **8 boards** $14 \div 3\frac{1}{2} = \frac{14}{1} \div \frac{7}{2} = \frac{14}{1} \times \frac{2}{7} = 4$, and 2 benches $\times$ 4 boards = 8 boards

Fraction Calculator Operations and Grids (page 175)

A.
1. 15
2. 6
3. $19\frac{1}{4}$
4. $5\frac{3}{4}$
5. $\frac{3}{4}$
6. $7\frac{1}{9}$
7. $3\frac{3}{32}$
8. $51\frac{1}{2}$

B.
9. **450** Half of 500 miles is 250 miles. Solve to find the total: $250 + \frac{4}{5}(250) = 250 + 200 = 450$ mi.

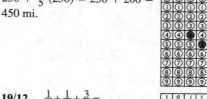

10. **19/12** $\frac{1}{3} + \frac{1}{2} + \frac{3}{4} = \frac{4}{12} + \frac{6}{12} + \frac{9}{12} = \frac{19}{12}$

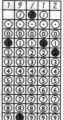

11. **300** $\$8 \times 7\frac{1}{2} \times 5 = \$300$

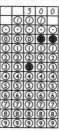

12. **3/2** To double the amount of sugar, multiply by 2: $\frac{3}{4} \times 2 = \frac{3}{4} \times \frac{2}{1} = \frac{6}{4} = \frac{3}{2}$

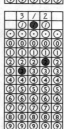

Ratios and Rates (page 176)

A.
1. $\frac{6}{1}$
2. $\frac{5}{6}$
3. $\frac{5}{9}$
4. $\frac{3}{11}$
5. $\frac{14}{5}$
6. $\frac{9}{1}$
7. $\frac{2}{1}$
8. $\frac{2}{3}$
9. $\frac{1}{10}$

B.
10. 125 sheets per $1
11. 20 miles per gallon
12. $0.43 per ounce
13. 100 square feet per quart
14. 0.25 pound per cup
15. 40 students per bus
16. 3 liters/2 quarts or 3:2 or 3 to 2
17. $\frac{3}{5}$ or 3:5 or 3 to 5
18. $15.50 per hour
19. $0.32 per orange
20. $\frac{11}{25}$ or 11:25 or 11 to 25
21. $\frac{2}{5}$ or 2:5 or 2 to 5
22. $\frac{10}{31}$ or 10:31 or 10 to 31
23. 16 ounces for $2.40 = $0.15 per ounce. 24 ounces for $3.48 = **$0.145 per ounce**—the better buy
24. 2 rolls for $2.45 = $1.225 per roll
3 rolls for $3.36 = $1.12 per roll
6 rolls for $5.88 = **$0.98 per roll**—the best buy

Proportions (page 177)

A.
1. $x = 12$
2. $x = 15$
3. $x = 20$
4. $x = 9$
5. $x = 7.5$
6. $x = 3$
7. $x = 10$
8. $x = 8$
9. $x = 2.1$
10. $x = 2$
11. $x = 88$
12. $x = 0.8$

B.
13. **150 points** $\frac{4}{60} = \frac{10}{x}$, and $60 \times 10 \div 4 = 150$
14. **300 computers** $\frac{20}{100} = \frac{x}{1,500}$, and $20 \times 1,500 \div 100 = 300$
15. **$16.50** $5.50 \times 3 = \$16.50$
16. **15 feet** $\frac{3 \text{ in.}}{9 \text{ ft.}} = \frac{5 \text{ in.}}{x}$, and $9 \times 5 \div 3 = 15$
17. **12.8 ounces** $\frac{1}{10} = \frac{x}{128}$, and $1 \times 128 \div 10 = 12.8$
18. **11.25 hours** $\frac{2 \text{ in.}}{3 \text{ hr.}} = \frac{7.5 \text{ in.}}{x \text{ hr.}}$, and $3 \times 7.5 \div 2 = 11.25$

Percents (page 178)

A.
1. 12.6
2. 10%
3. 40
4. $7.20
5. 70%
6. 37.5
7. 120
8. 25%

B.
9. $2.08
10. 20%
11. 30,900
12. $117.30
13. 25%
14. 500 tickets

C.
15. $345
16. price after first discount: $445
price after second discount: **$333.75**
17. 20%
18. $679.15
19. 12.5%

Proportions and Percents with Calculators and Grids (page 179)

A.
1. 37.5%
2. 6.25%
3. $16,600
4. 120
5. 312.5
6. 57
7. $1.50
8. 400
9. 5%
10. $1.425, or $1.43 rounded to the nearest cent

B. 11.

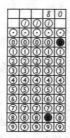

12.

13.

14.

Standard and Metric Measurement (page 180)

A.
1. 9 feet
2. 40 ounces
3. 3 cups
4. 1.5 feet
5. 2.25 tons
6. 5 minutes
7. 4 cups
8. 2 meters
9. 1,200 grams
10. 0.35 liter
11. 200 meters
12. 1.5 grams
13. 0.8 kiloliter

B.
14. 8 lb. 2 oz.
15. 2 yd. 1 ft.
16. 48 minutes 36 seconds
17. 24 quarts
18. 2,200 meters = 2.2 kilometers
19. 432 milligrams
20. 60.2 liters
21. 5.75 kilograms

C.
22. 2 hr. 20 min.
23. 6 ft. 3 in.
24. 20 packages
25. 60 grams

Perimeter, Area, and Volume (page 181)

A.
1. $P = 60$ inches
$A = 225$ square inches
2. $P = 60$ centimeters
$A = 216$ square centimeters
3. $P = 11.4$ feet
$A = 7.56$ square feet

B.
4. $V = 324$ cubic inches
5. $V = 46$ cubic centimeters
6. $V = 27$ cubic feet
7. $P = 20$ feet
8. 24 blocks
9. $P = 96$ inches
10. $V = 80$ cubic feet
11. $V = 420$ cubic inches
12. $A = 288$ square feet for 1 wall
288 square feet $\times$ 4 walls =
1,152 square feet

Irregular Figures (page 182)

Note: You may divide the figures differently.
1. $V = 6 \times 1.5 \times 1 + 5 \times 1.5 \times 1 = 9 + 7.5 =$ **16.5 cubic feet**
2. $P = 9 + 16 + 24 + 8 + 12 + 4 =$ **73 feet**
3. $A = 4 \times 12 + 12 \times 24 + 4 \times 9 = 48 + 288 + 36 =$ **372 square feet**
4. $P = 2 \times 6 + 2 \times 15 + 25 = 12 + 30 + 25 =$ **67 feet**
5. $12 \times 18 - 10 \times 16 = 216 - 160 =$ **56 square meters**

6. $A = 8 \times 2 + 6 \times 1 + 6 \times 1 = 16 + 6 + 6 =$ **28 square feet**

7. 12 cubic feet First convert measurements to feet: 12 in. = 1 ft., and 18 in. = 1.5 ft. $8 (1 \times 1 \times 1.5) = 12$ cu. ft.

GED Formulas (page 183)

A. **1. 175 miles** $d = rt = 70(2.5) = 175$ mi.

 2. 90,000 sq. ft. $A = s^2 = 300^2 = 90,000$ sq. ft.

 3. 67.41 square meters $A = bh = 10.7(6.3) = 67.41$ square meters

 4. 1,152 cu. ft. $V = lwh = 12(12)(8) = 1,152$ cu. ft.

 5. 19.625 square meters $A = \pi r^2 = 3.14(2.5)^2 = 19.625$ square meters

 6. 30 in. $P = s_1 + s_2 + s_3 = 10 + 10 + 10 = 30$ in.

 7. 37.68 cu. ft. $V = \pi r^2 h = 3.14(4)(3) = 37.68$ cu. ft.

 8. $326.25 $i = prt = \$1,500(.0725)(3) = \$326.25$

B. **9.** (1) 20×45

 10. (5) $\dfrac{2(18) + 2(24)}{12}$

 11. (3) $\$1.50 \times 5$

 12. (4) $3.14 \times 6^2 \times 15$

Tables (page 184)

 1. $233 $\$898 - \$665 = \$233$

 2. $8,796 $\$733 \times 12 = \$8,796$

 3. $151,740 $\$843 \times 12 \times 15 = \$151,740$

 4. 12% $\dfrac{560}{4,840} = .1157 = 11.57\%$, which rounds to 12%

 5. 69% $8,188 - 4,840 = 3,348$, and $3,348/4,840 = .6917 = 69.17\%$, which rounds to 69%

 6. Asia

 7. 1,700 $17 \times 100 = 1,700$

 8. 300 $3 \times 100 = 300$

 9. 3,000 $30 \times 100 = 3,000$

 10. $3,300 $\$22,000 \times .15 = \$3,300$

 11. $5,700 $\$38,000 \times .15 = \$5,700$

 12. $8,400 $\$33,050 - \$24,650 = \$8,400$

Graphs (page 185)

 1. Sunday

 2. Tuesday and Friday

 3. 6 degrees

 4. 3rd quarter

 5. 75%

 6. $31,500

 7. Jose and Albert

 8. $18,000

 9. 11.11% rounds to 11%

 10. Turner

 11. 17.24% rounds to 17%

 12. Lincoln

Probability (page 186)

1. $\dfrac{2}{5}$ **5.** 70%

2. $\dfrac{4}{7}$ **6.** 40%

3. $\dfrac{4}{7}$ **7.** $66\dfrac{2}{3}\%$

4. $\dfrac{1}{5}$

Data Analysis with the Calculator and Grids (page 187)

A. **1.** 5.8

 2. 33.67

 3. 17.08

 4. $27.708 rounds to $27.71

B. **5.** 9.5

 6. 213

 7. 6.8

 8. $2,335

C. **9.**

 11.

 10.

 12.

Math Handbook
Using the CASIO *fx*-260

When you take the GED Math Test, you will be given the CASIO *fx*-260 scientific calculator to use for Part I of the test. Calculators will then be collected before you start Part II. Although you can use any calculator to do the math in this book, you should practice using a scientific calculator before you take the GED Math Test.

Scientific calculators are useful because they automatically follow the order of operations. Unfortunately, having so many keys can be confusing. Study the diagram below to find the location of the keys you will need to do the work in this book. You will learn to use other calculator functions as you progress in your study of math.

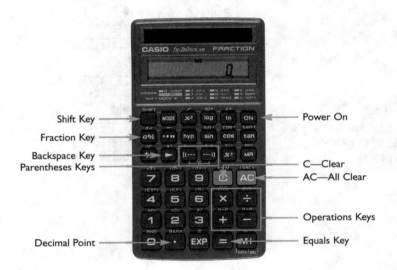

Shift Key
Fraction Key
Backspace Key
Parentheses Keys
Decimal Point

Power On
C—Clear
AC—All Clear
Operations Keys
Equals Key

If the display is blank, press [ON] to power up the calculator.

You should always see the letters DEG in the display window when you begin a calculation. If you see other letters, press [AC] to reset before you perform a calculation.

Basic Operations

The CASIO *fx*-260 follows the order of operations. In other words, if you enter a series of operations, the calculator will multiply and divide first and then add and subtract. If you want the calculator to perform the operations in some other order, you need to use parentheses.

EXAMPLE: 55 − 20 × 2
Enter: [5] [5] [−] [2] [0] [×] [2] [=]
| 15. |

EXAMPLE: (55 − 20) × 2
Enter: [(-] [5] [5] [−] [2] [0] [-)] [×] [2] [=]
| 70. |

In the first example, the calculator multiplied first and then subtracted, following the order of operations. In the second example, the parentheses told the calculator to subtract first and then multiply.

Special Keys

If you make a mistake while entering a calculation, you can press:

[►] to backspace and delete digits
[C] to clear the last operation performed
[AC] or [ON] to clear everything and start over

Using the Fraction Keys

As you know, fractions and decimals are two ways to express parts of a whole. Ordinary calculators cannot work problems with fractions unless the fractions are first converted to decimals. Scientific calculators have special fraction keys that allow you to enter fractions as they are normally written. These keys can save you valuable time when you take the GED Math Test.

Study the diagram below to find the fraction keys on the CASIO *fx*-260. You will need to use the $\boxed{\text{SHIFT}}$ key to use one of the fraction keys.

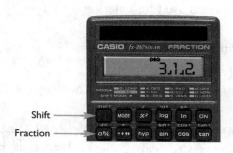

The display shows the mixed number $3\frac{1}{2}$.

The whole number, numerator, and denominator are separated by the symbol ⌐.

To enter a fraction or mixed number, press $\boxed{a^b/c}$ between each part.

EXAMPLES: To enter $\frac{4}{5}$, press $\boxed{4}$ $\boxed{a^b/c}$ $\boxed{5}$.

To enter $5\frac{1}{3}$, press $\boxed{5}$ $\boxed{a^b/c}$ $\boxed{1}$ $\boxed{a^b/c}$ $\boxed{3}$.

Use the $\boxed{=}$ key to reduce a fraction to lowest terms.

EXAMPLE: Reduce $\frac{48}{64}$ to lowest terms.

Press: $\boxed{4}$ $\boxed{8}$ $\boxed{a^b/c}$ $\boxed{6}$ $\boxed{4}$ $\boxed{=}$ **Display:** | 3⌐4. |

Answer: The fraction $\frac{48}{64}$ reduces to $\frac{3}{4}$.

Use the calculator to perform operations with fractions. If the answer is a mixed number, you can change it to an improper fraction by pressing $\boxed{\text{SHIFT}}$ $\boxed{a^b/c}$ to access the $\boxed{d/c}$ function.

EXAMPLE: $\frac{5}{6} + 3\frac{2}{3}$

Press: $\boxed{5}$ $\boxed{a^b/c}$ $\boxed{6}$ $\boxed{+}$ $\boxed{3}$ $\boxed{a^b/c}$ $\boxed{2}$ $\boxed{a^b/c}$ $\boxed{3}$ $\boxed{=}$ **Display:** | 4⌐1⌐2. |

Change the mixed number to an improper fraction.
Press: $\boxed{\text{SHIFT}}$ $\boxed{d/c}$ **Display:** | 9⌐2. |

Answer: The answer is $4\frac{1}{2}$ or $\frac{9}{2}$.

How to Grid Answers

On the GED Math Test, you will enter some of your answers on number grids. When the machine scores your answer sheet, it will score only the bubbles that you filled in. You need to learn how to correctly use these grids to make sure you get full credit for your work.

Look at the blank grid shown to the right. Locate the following features:

- The first row has blank boxes. Use this row to write your answer, fitting one number or symbol to each box as needed.
- The second row has slash marks used for entering fractions.
- The third row has decimal points used for entering decimal numbers.
- The remaining rows show the digits 0 through 9.

Study the following examples to learn the rules for filling in number grids.

RULE 1: You can start your answer in any box as long as it fits within the grid. If a column is not needed, leave it blank.

A theater has 1,200 seats, but on opening night, only 846 seats were filled. How many seats were empty?

Answer: 354 empty seats

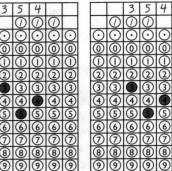

RULE 2: If the answer is not a whole number, enter it as either a fraction or a decimal.

A bag has $\frac{3}{8}$ pound of nails. If you remove $\frac{1}{8}$ pound, how many pounds are left?

Answer: $\frac{1}{4}$ or **0.25 pound**

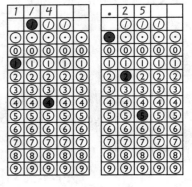

RULE 3: If the answer to a problem is a mixed number, enter it as an improper fraction or change it to a decimal.

What is the sum of $1\frac{3}{5}$ and $\frac{9}{10}$?

Answer: The sum is $2\frac{1}{2}$, but you must enter it as $\frac{5}{2}$ or **2.5.**

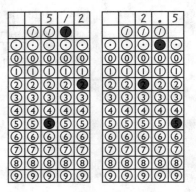

Common Equivalencies

As you prepare to take the GED Math Test, you will find it helpful to memorize the basic equivalencies shown on this page.

Common Fractions, Decimals, and Percents		
$\frac{1}{2}$	0.5	50%
$\frac{1}{3}$	about 0.33	$33\frac{1}{3}\%$
$\frac{2}{3}$	about 0.67	$66\frac{2}{3}\%$
$\frac{1}{4}$	0.25	25%
$\frac{3}{4}$	0.75	75%
$\frac{1}{5}$	0.2	20%
$\frac{2}{5}$	0.4	40%
$\frac{3}{5}$	0.6	60%
$\frac{4}{5}$	0.8	80%
$\frac{1}{10}$	0.1	10%
$\frac{3}{10}$	0.3	30%
$\frac{7}{10}$	0.7	70%
$\frac{9}{10}$	0.9	90%

Parts of a whole can be written as fractions, decimals, or percents. On the GED Math Test, use the form that is easiest for you to solve the problem.

English Measurement Equivalencies

LENGTH
1 foot (ft.) = 12 inches (in.)
1 yard (yd.) = 3 feet = 36 inches
1 mile (mi.) = 5,280 feet

TIME
1 minute (min.) = 60 seconds (sec.)
1 hour (hr.) = 60 minutes
1 day = 24 hours
1 week (wk.) = 7 days
1 year (yr.) = 365 days

VOLUME
1 cup (c.) = 8 fluid ounces (fl. oz.)
1 pint (pt.) = 2 cups
1 quart (qt.) = 2 pints = 4 cups
1 gallon (gal.) = 4 quarts

WEIGHT
1 pound (lb.) = 16 ounces (oz.)
1 ton = 2,000 pounds

Metric Measurement Equivalencies

LENGTH
1 meter (m) = 1,000 millimeters (mm)
1 meter = 100 centimeters (cm)
1 centimeter = 10 millimeters
1 kilometer (km) = 1,000 meters

VOLUME
1 liter (L) = 1,000 milliliters (mL)

WEIGHT
1 gram (g) = 1,000 milligrams (mg)
1 kilogram (kg) = 1,000 grams

Formulas Page

This page will be provided when you take the GED Math Test. The highlighted formulas will be helpful as you do the work in this book.

FORMULAS

AREA of a:

square	Area = side^2
rectangle	Area = length × width
parallelogram	Area = base × height
triangle	Area = $\frac{1}{2}$ × base × height
trapezoid	Area = $\frac{1}{2}$ × (base$_1$ + base$_2$) × height
circle	Area = π × radius^2; π is approximately equal to 3.14.

PERIMETER of a:

square	Perimeter = 4 × side
rectangle	Perimeter = 2 × length + 2 × width
triangle	Perimeter = side$_1$ + side$_2$ + side$_3$

CIRCUMFERENCE of a:

circle	Circumference = π × diameter; π is approximately equal to 3.14.

VOLUME of a:

cube	Volume = edge^3
rectangular solid	Volume = length × width × height
square pyramid	Volume = $\frac{1}{3}$ × (base edge)^2 × height
cylinder	Volume = π × radius^2 × height; π is approximately equal to 3.14.
cone	Volume = $\frac{1}{3}$ × π × radius^2 × height; π is approximately equal to 3.14.

COORDINATE GEOMETRY

distance between points = $\sqrt{(x_2 - x_1)^2 + (y_2 - y_1)^2}$; (x_1,y_1) and (x_2,y_2) are two points in a plane.

slope of a line = $\frac{y_2 - y_1}{x_2 - x_1}$; (x_1,y_1) and (x_2,y_2) are two points on the line.

PYTHAGOREAN RELATIONSHIP

$a^2 + b^2 = c^2$; a and b are legs and c the hypotenuse of a right triangle.

MEASURES OF CENTRAL TENDENCY

mean = $\frac{x_1 + x_2 + ... + x_n}{n}$, where the x's are the values for which a mean is desired, and n is the total number of values for x.

median = the middle value of an odd number of <u>ordered</u> scores, and halfway between the two middle values of an even number of <u>ordered</u> scores.

SIMPLE INTEREST   interest = principal × rate × time
DISTANCE   distance = rate × time
TOTAL COST   total cost = (number of units) × (price per unit)

Reprinted with permission from the GED Testing Service. ©2001, GEDTS